Fundamental Ecology

Intext Series in ECOLOGY

ARTHUR S. BOUGHEY, *Editor*
University of California, Irvine

Fundamental Ecology

Arthur S. Boughey
University of California, Irvine

Intext Educational Publishers
College Division of Intext
Scranton   San Francisco   Toronto   London

ISBN 0–7002–2363–0
Library of Congress Catalog Card No.: 77–151647

Copyright ©, 1971, International Textbook Company

Second Printing, September, 1972

Series Preface

As we move into the decade of the 70's we are confronted with dire threats of imminent environmental disaster. While prophecies as to the actual doomsday vary from five years to thirty years from now, no professional ecologist seems willing to state categorically that mankind will survive into the next millenium unchallenged by any ecocatastrophe. Some indeed believe that before this time we and most of our familiar ecosystems are inevitably doomed to extinction.

Enough has now been said and written about such predicted disasters to instill in students, governments and the public at large an uneasy feeling that something may be amiss. Terms such as *pollution, natural increase* and *re-cycling* have begun to assume a realistic and more personal note as the air over our cities darkens, our rivers are turned into lifeless fire-hazards, our domestic water becomes undrinkable, and we have to stand in line for any form of service or amenity.

Politicians, scientists and the public have responded variously to this new situation. Tokenism is rampant in thought, word and deed. Well-intentioned eco-activist groups have mushroomed, not only among youth, who are the most threatened as well as the most understanding segment of our societies. More specifically, in the restricted field of college texts, appropriate ecological chapters have been hurriedly added to revised editions. No biological work is now permitted to conclude without some reference to human ecology and environmental crises.

The purpose of this new ecological series is to survey without undue overlap the major fields of our present environmental confrontation at an introductory college level. The base text for the series presents an oversight of the ecological fundamentals which are relevant to each issue. In association with the works listed in its bibliographical references, it can stand alone as a required text for an introductory college course. For such use each chapter has been provided with a set of review questions. For more extensive courses, the base text leads into each series volume, and the particular area of environmental problems which this explores.

This series treats, subject by subject, the main points of impact in this current ecological confrontation between man and his environment. It presents in breadth and in depth the problems of pollution, pesticides, waste materials, population control, and the resource exploitation which imminently threaten to overwhelm us. Each volume in the series is a definitive study prepared by a specialist in the field, writing from an intimate personal experience of his area, relating but not overlapping his subject with other volumes in the series. Uniquely assembled in each volume will be information which presently is not available without extensive bibliographical research, at the same time arranged and interpreted in a more readily assimilable form. Extensive illustrative material, much of it original, still further facilitates a ready comprehension of the matter presented.

This is an exciting series. The urgency and ferment which have been experienced by all those associated with it cannot fail to be transmitted to the reader. The series confounds the prophets of doom, for it illustrates that given a proper understanding of ourselves and our ecological world, there is yet time for action. This time may be short, but sufficient if we exercise now the characteristics of courage and resolution in which, at times of great crisis, our species has never previously been found wanting.

Preface

Concident with the birth of this current decade a new and ominous-sounding phrase began to ring in our ears, *environmental crisis*. Newspapers and popular magazines, radio and television commentators, politicians and administrators alike find the phrase an essential addition to any public presentation. Despite this extensive exposure little is ever added either to emphasize the real meaning of this crisis or to define its proportions. There has even been a tendency to dissociate it from its underlying causes, and especially from the most basic of all, the *contemporary population explosion*.

This current series on the ecological origins of our environmental crisis is designed to fill this information gap, and to present both a comprehensive review of its various aspects and an authoritative statement of its causes. The basic text that accompanies this series on Ecology was prepared so as to avoid the repetition in every volume of fundamental ecological concepts essential to an understanding of each aspect of the crisis. In the basic text the discipline of ecology is reviewed in simple terms readily comprehensible to readers with no greater technical knowledge than can be acquired from a high school survey course in biology. This text is therefore suitable for use by beginning college students and by any other interested enquirer.

There already exist a number of elementary texts in ecology. Invariably, however, these have two features that reduce their suitability as basic texts when considering the environmental crisis. First, their emphasis is either on natural history or field biology because the popular approach to ecology has traditionally been from these directions. Second, and largely because of this emphasis, such reference as is made to environmental crises primarily relates to the preservation of the countryside and the conservation of natural communities.

These are extremely worthy objectives, but our present human societies comprise essentially aggressive and exploitive *urban* cultures. All

aspects of our present environmental crisis are derived directly or indirectly from the apparently insatiable demands of this urban civilization.
The solutions for each component crisis must therefore be sought in the
cities, which is where the predicted *ecocatastrophes* will first befall. Urban
systems are the foundations upon which we have to construct the rest of
our world as best we can, as we struggle to achieve stability and maintain
diversity in such plundered communities as still remain.

Ecology is variously defined. When it is regarded as a study of communities in relation to their environment, a survey of the principles and
concepts which have developed within the discipline reveals why it is
necessary to maintain stability and diversity. The present text emphasizes
these requirements and the relationship between fundamental ecological
principles and each individual crisis. It identifies the areas that are provided with a more extensive treatment in the individual volumes in the
series.

The ecologist has invaded natural communities, has observed them,
classified them, manipulated them experimentally and analyzed them,
and produced a series of ecological principles and concepts which attempt
to explain their distribution, appearance, structure and function. In very
much the same way ecology must now invade our urban communities,
penetrating into the industrial complexes, the ghettos, the affluent suburbs, the schools, the recreation and the health centers, and even more
intimate aspects of urban behavior and culture. The primary object of
this text is to indicate the extent to which our present general ecological
knowledge is applicable to such an endeavor, as well as to introduce the
individual environmental problems included in the series.

After this urban exposure, ecology will never again be the same. Nor
will human life. The issues may be simplistically reduced to the single
question of survival; indeed, ecology is now hailed as the *science of survival*. However sophisticated our society and affluent our way of life, if
we are to survive we can no more defy ecological laws than we can ignore
the law of gravity. As with the sensation of gravity in sky-diving, we may
not feel ourselves falling on the way down, but this in no way diminishes
the impact of the terminal ecocatastrophic crunch. To continue the simile,
for many environmental situations the main chute has already carried
away or failed to open. If we are to avoid utter disaster, we must hasten
to deploy the emergency chutes without a moment's delay.

Arthur S. Boughey

University of California, Irvine
May, 1971

Contents

Introduction

For several centuries we have avidly seized every opportunity for the industrial exploitation of contemporary scientific inventions. At the same time we have contrived to ignore or to minimize the ominous implications for our own species of many fundamental observations and discoveries. The triumph of *mind* over *matter* which our technocracy believed it had achieved was generally thought to confer on our own species some special exemption from the biological laws that controlled the rest of the living world. Greek philosophers like Aristotle and Plato, on whose precepts our present Western civilization was based, showed a greater realism. They appreciated that there was an optimum size for their cities and a limit to the productivity of their lands. To such possible restrictions on unfettered development, Renaissance and Industrial man gave no thought. Engrossed with the rapidly increasing power of human technology, we had ceased to reflect on such unwelcome possibilities as an eventual limitation of our burgeoning urban expansion.

Effects of the Industrial Revolution

This disregard of resource limitations was of little consequence in the earlier phases of the Industrial Revolution. Farming, a pursuit in which a majority of the population was still engaged, was largely unmechanized; the animal power used in agricultural operations was maintained on products which had been locally grown. Waste products from the draft animals were returned to their parent soils and recycled in the same localities in which they originated. Cities were not yet so large or so numerous that they seriously interrupted such local recycling processes.

This is still essentially the condition of the underdeveloped nations of the world which contain two-thirds of the human population. Advanced industrial societies by contrast have now concentrated their

1

populations in cities. They are exploiting fossil fuels, minerals, and other nonrenewable resources obtained from outside these urban assemblages at a rate which it will soon prove difficult to sustain and quite impossible even temporarily to expand to meet ever-growing needs. Waste products, instead of being recycled, are converted into nonbiodegradable substances which accumulate about the cities at alarming rates embarrassing to both local and national authorities.

Failure to return these various used materials into the recycling systems has been associated with the release of other wastes and by-products into the air sheds on which we rely for oxygen and the watersheds which meet our vital water needs. Till now we had regarded these air and water systems as common self-cleansing pools freely available for exploitation, and had given no thought as to their possible depletion or destruction. The cumulative results of our neglect of such fundamental ecological considerations are explained in this text and further described in the various volumes in this series.

The Development of Ecology

While various environmental crises have thus steadily been enveloping us as a result of our ever-increasing level of urbanization, the discipline of ecology has independently come to maturity. It has emerged as an integrated study of *populations, communities,* and *ecosystems.* These three entities represent the highest levels of biological organization in an ascending series commencing with intracellular structures at the molecular level and proceeding upward through tissues and organs to individual organisms. *Populations* can be regarded as groups of organisms having a common origin, form and function; they are generally treated as identical with *species. Communities* are associations of populations linked by some interdependent function. *Ecosystems* are conceptual systems formed by relating a community or communities with the totality of prevailing environmental factors.

Despite this integration of the higher levels of biological organizations, ecologists generally still pay scant attention to the structure of human populations and the functions of urban communities. There are several reasons for this apparent dissociation, as will be discussed. These go beyond our unwillingness to admit that our own species is also subject to ecological laws. The unique cultural power we have developed now permits us to modify our environment almost at will. This may well be the primary source of our belief that the seemingly infinite resourcefulness of our minds and our technology permit us to ignore fundamental ecological considerations.

Changing Emphasis in Ecology

The discipline of ecology in its latest form as the study of living systems in relation to one another and to their environment, is of recent origin. Its beginnings date no earlier than the second half of the nineteenth century, when explorations of the natural history of the earth were undertaken on an impressive scale by the expansionist societies of the Western world. The discovery of many different forms of life unfamiliar to Western scientists, combined with the introduction of the experimental approach into the once largely observational and anecdotal study of natural history, caused the new science to develop vigorously. At first it tended to do so along two divergent lines. Animal ecologists were concerned more especially with *population dynamics*, the experimental study of the environmental factors which determine the size and cause of fluctuations in the numbers of individual species under natural conditions, sometimes known as *autecology*. Plant ecologists remained for a number of years involved in inventorial activities, observing and analyzing the range of communities which they encountered, and attempting to correlate particular forms and structures with physical and chemical factors of the macroenvironment, a subject sometimes described as *synecology*.

These early years saw the development of many basic ecological concepts and were attended by a growing excitement as these were tested and found to define many natural and experimental situations. Ecology began to have a marked influence on what before had been empirical management sciences such as agriculture, horticulture, and forestry. Autecological studies soon produced an increased appreciation of the great complexity of the environment. Critical explorations of the basis for the regulation of animal populations still further emphasized this and directed attention toward the nature of competition between species populations. At the same time a growing realization of the significance of gene flow in populations resulted in the application of genetical principles to theoretical ecology and the "niche" concept was evolved. An *ecological niche* is defined as the unique complex of factors which characterizes the environment of any given population. Armed with this concept, it was possible to launch out into a consideration of species diversity on the one hand and the coevolution of many mutualistic relationships between populations on the other.

Meanwhile it became apparent that populations were associated in communities by links other than evolutionary ones. Concepts of food chains and foodwebs developed as hypotheses by which to delimit the many processes of energy, matter, and nutrient exchange which occurred in these communities. A major contribution synecology made at this time was to stress the continuity rather than the discreteness of ecological entities, whether these were populations and communities or chemical substances and physical factors. Synecological studies began to stress also the delicate balance of nature, which was found to be much more fragile and more easily disturbed than had hitherto been imagined.

As ecological experimentation passed into a fully quantified phase, mathematical modeling and analysis received more emphasis, and directed greater attention toward the physical aspects of the discipline. Ecology had finally ceased to be basically a field study of natural history, and instead now centered around theoretical, laboratory, and field research on populations and simple experimental communities. Inevitably the emphasis on interactions with physical factors became linked with the newer synecological community concepts of biological continua and gradients. The interplay between living organisms, materials, and energy began to be stressed. Populations were seen to react with one another not only within a biological system but also within an interdependent physical one. Some twenty-five years ago the term *ecosystem* was coined and adopted to describe this complex system formed by biological communities and the physical environment. The concept of the ecosystem finally reunited autecological and synecological studies in a mathematically definable relationship in which biological reactions could be described in terms of physical laws. What finally established the ecosystem concept as the fundamental basic unit of ecology was the general adoption by ecologists of the technique of systems analysis which had been developed more especially by engineers.

The Current Definition of Ecology

The introduction of the *systems approach* to ecology scarcely a decade ago, together with an appreciation of the significance of the ecosystem concept, has brought ecology into its current phase. It is now possible to define ecology as Odum (1959) has done, simply as *the study of ecosystems.*

Defined in this manner the scope of ecology is immense. Although it is possible in a general way to survey the whole field, usually it has to be subdivided for more detailed considerations. The older partition into plant and animal ecology is no longer considered valid, nor is the later variant of this into synecology and autecology. Ecology now loosely separates into such areas as population ecology, community ecology, evolutionary ecology, environmental ecology, behavioral ecology, mathematical ecology, marine ecology and human ecology. For the purpose of this introductory text, such a partitioning is unnecessary, and the principles and concepts of ecology will be viewed as a whole. This has become even more desirable with the general adoption by ecologists of the systems approach.

The Systems Approach

The *systems approach* has slightly varying meanings in the differing contexts of particular disciplines. For the ecologist it signifies more particularly the subdivision of ecological processes into simple unitary stages. This analysis permits the simulation of even very complex ecological situations in computer models by iterative logical procedures (Van Dyne, 1966). A diagrammatic representation of this process is illustrated in Fig. Intro.-1.

In terms of specific ecological investigations, the systems approach is now increasingly used in combination with two technological developments, continuous electronic telemetering and automated data processing by means of analog or digital computers. This make possible for example the construction of a single model which may be applied to as widely varying ecological phenomena as the development of particular leaf shapes, the form of the home range of a rodent, the nature of feeding positions in bird flocks and the formation of wing patterns in insects (Bridges, 1970). The possibilities for continuous telemetering have been enormously advanced by technological spin-off from the several space exploration projects, and the development of various satellite sensing devices for this program. The full impact of these recent developments on ecology has still to be felt.

This adoption of systems analysis has contributed to a radical change in the direction and scope of ecological studies. The previous essentially experimental approach involved the investigation of one isolated population, and the manipulation of one specific factor while all others were held steady. Now it has become possible to investigate the structure and functions of populations, communities, and ecosystems under natural conditions without the experimental manipulation previously necessary in order to reduce the scale of such observations to a manageable size. This change in approach is perhaps best exemplified in *microbial ecology* where autecology had reached its most extreme form with laboratory studies on pure cultures of microorganisms (Smith, 1970). The introduction of the systems approach emphasizes the essentially holistic view of ecology (Smuts, 1926) which was incorporated in the ecosystem concept. In this sense ecology as it is now understood is the study of populations and communities as a whole in relation to one another and to their total environment.

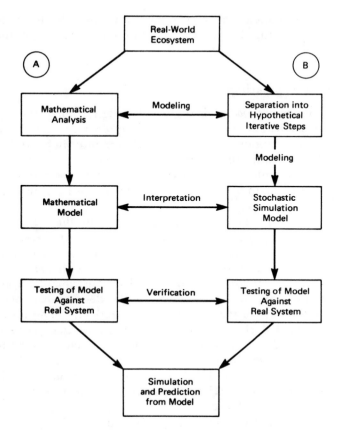

Fig. Intro.-1. The systems approach to ecosystems. The systems ecologist constructs either mathematical (A), or computer (B), models to simulate the various functions and activities of the ecosystems he studies. Even simple ecosystems have such complicated transfer function matrices that only computer models can realistically be employed for their simulation. The sequence of *modeling, interpretation,* and *verification* is extensively applied by mathematicians, physicists, and engineers, but has only recently been introduced to ecology.

Ecology and the Environmental Crisis

The purpose of this base text is to identify the various fundamental principles included in this new discipline of ecology and to explain them in elementary language. As each ecological concept is defined it is related to the enviromental confrontations which have developed as a consequence of ignoring its requirements.

We are now finally coming to appreciate that we are limited by the resources of water, air, energy, and minerals which we have within this solar system, and more particularly upon our own planet. We are beginning to take into consideration the whole of our global population and its total utilization of our planetary resources. We are coming to realize that in many cases we are already dangerously close to the maximum extension of this resource utilization. In a few instances it is apparent that we have indeed exceeded it.

This base text therefore explains the fundamental ecological processes to which we must conform, the various volumes in the series define the limits our developing civilization must observe. When we understand and appreciate the significance of these controls and these limitations, we shall have taken a great step forward in understanding the actions our global populations must now undertake in order to ensure the continuing survival and progress of our contemporary cultures.

The environmental crisis may be conveniently described under such headings as air pollution, water pollution, population control, pesticides, and conservation. Each of these topics is dealt with specifically in an individual volume of this series. Those who are interested in doing so may therefore proceed from the origins of each crisis into a more detailed study of the subject. Likewise it is possible to reverse this procedure and progress from the study of a particular environmental problem to its fundamental origins. For those who have neither the time or inclination to pursue each problem to this extent, selected readings are cited in the human ecology section of the bibliography at the end of each chapter which provide a shorter survey of each major crisis. The bibliographies also provide references to the source material from which much of the fundamental information contained in this work is drawn.

Ecological Literature

For English-speaking readers, ecological source material is provided especially in two periodicals. One is *Ecology* the official journal of the Ecological Society of America, which also produces *Ecological Monographs* for the publication of more lengthy papers. The other is *The Journal of Ecology* issued by the British Ecological Society which has two other regular publications, *The Journal of Animal Ecology* and *The Journal of Applied Ecology*. Research on aquatic ecosystems is often published in *Limnology and Oceanography*. Review features, and the introductions to many of the papers in these journals, provide instructive reading for beginning students of ecology. The same applies to a number of the "in-depth" papers appearing in the journals *Advances in Ecological*

Research and *The American Naturalist.* Review articles and statements also appear in two general science American publications *Science* and *Bioscience.* At a more popular science level the periodicals *New Scientist, Natural History* and *The Scientific American* contain many features surveying recent ecological developments in nonspecialist language.

Literature on the various environmental crises is exceedingly scattered, as is indicated when reference is made to the bibliographies provided at the end of each chapter. The magazine *Environment* covers many aspects of current environmental crises; again it is written in nonspecialist language. The *International Journal of Environmental Studies* includes the same field at a more specialist level. The *Population Bulletin* provides useful statistics, and *Daedalus* contains many population studies.

For those who wish to read somewhat further into the basic subjects, numerous readings volumes are now available. A number of anthologies which contain excellent papers dealing with many aspects of human ecology and environmental crises have also appeared. The bibliography lists examples of both kinds of collections.

As this book is designed for use as a college text, a list of suggested review questions is provided at the end of each chapter. These may be used as a comprehension test for the material which has been presented.

Conclusion

Our civilization is currently extremely attached to "in" words. In 1968 it was the turn of *explosion;* the "information explosion," the "population explosion," the "technological explosion," the "communications explosion." In 1969 the word was *confrontation;* in 1970 fate has decided it is to be *ecology.* This brief introductory chapter is intended to explain what currently is understood by the term *ecology,* and the range and scope of activities which this discipline presently encompasses.

Ecology is something we can all understand. The last decade belonged to the molecular biologists. Their work enabled quantum advances to be made in biological theory, and has led to great progress in the understanding of biological processes at all levels of organization. It emphasized the unity of life rather than its diversity and helped us appreciate that we are an integral part of the ecological systems we occupy. For nonspecialists, however, it was and still remains very difficult to grasp the full implications of this molecular work. The knowledge of chemistry, physics, and mathematics as well as biology required for a complete understanding of this subcellular discipline is beyond the reach of many students and the majority of the public. Some aspects of ecology likewise demand a wide knowledge of other disciplines or extensive mathematical experience, but the basic principles are more readily comprehensible. When we have the courage and resoulution to apply these to our own behavior, it can readily be understood how we have allowed ourselves to be overtaken by many environmental crises, and what we have to do to reduce or to avoid these in the future.

The various aspects of ecology that can now be identified, the ecosystem with its interwoven energy and material pathways, the growth and interaction of its populations, their evolution and behavior, their environmental relations and feedback mechanisms, the structure, functions, and development of communities, are now considered in succeeding chapters. Finally the text concludes with a brief survey of human origins and their relationship to the crises that have arisen as a result of our continuing failure to recognize the ecological limits which even our unique species cannot ignore without courting the risk of ultimate ecocatastrophe and inevitable extinction.

Bibliography

References

Adams, L. (ed.), *Population Ecology*. Belmont, Calif.: Dickenson, 1970.

Boughey, A. S. (ed.), *Population and Environmental Biology*. Belmont, Calif.: Dickenson, 1968.

——— (ed.), *Contemporary Readings in Ecology*. Belmont, Calif.: Dickenson, 1969.

Bridges, K. W., "The Quantitative Description of Leaf Shape," Ph.D. thesis, Univ. California Irvine, 1970.

Connell, J. H., D. B. Mertz, and W. W. Murdoch (eds.), *Readings in Ecology and Ecological Genetics*. New York: Harper, 1970.

Hazen, W. E. (ed.), *Readings in Population and Community Ecology* rev. ed. Philadelphia: Saunders, 1970.

Klopfer, P. H. (ed.), *Behavioral Ecology*, Belmont, Calif.: Dickenson, 1970.

Odum, E. P., and H. T. *Fundamentals of Ecology*, 2nd Ed. Philadelphia: Saunders, 1959.

Riley, H. P. (ed.), *Evolutionary Ecology*, Belmont, Calif.: Dickenson, 1970.

Smith, H., "Microbes and Society," *New Scientist* 45:112–113, 1970.

Smuts, J. C., *Holism and Evolution*. New York: Macmillan, 1926.

Van Dyne, G. M., "Ecosystems, Systems Ecology and Systems Ecologists," *Oak Ridge National Laboratory Report* 3957:1–31, 1966.

Further Readings in Human Ecology

Boughey, A. S., *Man and the Environment: an Introduction to Human Ecology and Evolution.* New York: Macmillan, 1971.

Bresler, J. B. (ed.), *Environments of Man.* Reading, Mass.: Addison-Wesley, 1968.

Costin, A. G., "Replaceable Resources and Land Use," *Journal of the Australian Institute of Agricultural Science,* 25:3–9, 1959.

Cox, G. W. (ed.), *Readings in Conservative Ecology.* New York: Appleton, 1969.

Darling, F. F., "Conservation and Ecological Theory," *J. of Ecology* 52 (suppl.): 39–45, 1964.

Detwyler, T. R. (ed.) *Man's Impact on Environment.* New York: McGraw Hill, 1971.

Hardin, G. (ed.), *Population, Evolution and Birth Control.* rev. ed. San Francisco: Freeman, 1969.

Shepard, P., and D. McKinley (eds.), *Subversive Science.* Boston: Houghton Mifflin, 1969.

Thant, U. "Human Environment and World Order," *Intern. J. Environ. Studies* 1:13–17, 1970.

White, L., "The Historical Roots of our Ecological Crisis," *Science* 155:1203–1207, 1967.

Young, L. B. (ed.), *Evolution of Man.* New York: Oxford University Press, 1970.

Review Questions

1. Discuss the nature and scope of ecology.

2. What are the essential features of the systems approach? To what aspects of ecology can systems analysis presently be most effectively applied?

3. Outline the range of our present environmental crises: Why is it necessary to approach these from an ecological point of view?

4. What do you understand by the term "human ecology"?

5. Discuss the changes which have taken place in ecology during the last decade.

6. If you were engaged in the preliminary search for source material for a term paper on an ecological subject, name the scientific periodicals or journals whose recent numbers you would consult first. Provide a brief outline of the nature of the material appearing in each publication you mention.

1

Ecosystems

The introduction developed the idea of the ecosystem as the fundamental concept of ecology. We saw how this provides the essential link between the previously divergent studies of populations and communities. We saw also how the central importance of the ecosystem has been still further emphasized by the application to ecology of the systems approach. Associating this with computer data processing has made possible the construction of predictive simulation models. These provide forecasts of resource needs and estimates of exploitation rates.

In this first chapter the ecosystem is reviewed in greater detail, together with certain dependent but equally important concepts including those of *trophic levels, productivity, energy* and *nutrient budgets, ecological pyramids* and *biogeochemical cycling.* Some of these emerge from considerations of ecosystem *structure,* others from examining ecosystem *function,* aspects which it is not always possible to separate clearly.

An understanding of the significant implications of the ecosystem concept is best developed by starting with the axiom that living organisms on this earth cannot exist in physical and biological isolation. There have to be interactions between such organisms, and with the physical features and chemical substances of their environment. As defined here, the ecosystem is a complex conceptual unit composed of these living organisms and their environment, and characterized by dependent cause-effect pathways. It is thus basically an energy, material, and information transfer system in which various interactions provide feedback mechanisms controlling the numbers and development of its component organisms (Patten, 1959). The fundamental grouping of such organisms is the *population.* This has already been defined as a collection of similar organisms having a common origin, between whom no barrier exists preventing the exchange of reproductive material (if they possess the necessary mechanisms to achieve this exchange). Such groups of organisms are alternatively known as *species*. Within particular ecosystems, populations as so defined may be linked by energy exchange interactions into aggregations known as *communities*. *Populations* are therefore groups of organisms with *genetical origins* and *energy relationships* in common. *Communities* are groups of populations with interdependent *functional* relationships. *Ecosystems* are conceptual units within which communities exchange energy, materials and information with one another and with their physical environment (Fig. 1-1).

Basic Elements of Ecosystems

Any ecosystem must thus contain the following:
1. An abiotic element (the physical environment)
2. A biotic element (populations of autotrophic and/or heterotrophic organisms forming communities)
3. Energy input and utilization
4. Nutrient input and cycling

The relationship between these several characteristics of an ecosystem are illustrated schematically in Fig. 1-2.

The Abiotic Element

The abiotic element includes physical factors of the environment such as the temperature, moisture, light, and altitudinal parameters, together with chemical features of the habitat like the relative availability of essential nutrients, especially nitrates, phosphates, and potassium salts, and the levels of oxygen and carbon dioxide pressures. These various items of the abiotic element effect ecosystems in two ways, determining both their *nature* and their *function*. Particular physical and chemical features determine the *nature* of the ecosystem because they limit the range of organisms which will be represented in its populations. These also, together with other physical and chemical factors, control ecosystem *function* by determining the *rate* at which population interactions may occur. Both these aspects of the influence of the abiotic element on ecosystems are considered further in the next chapter. They are of primary importance in relation to the problems presented in all the volumes of this series.

The Biotic Element

The biotic element of the ecosystem, its populations of living animals, plants, and microbes together form the *communities* of the eco-

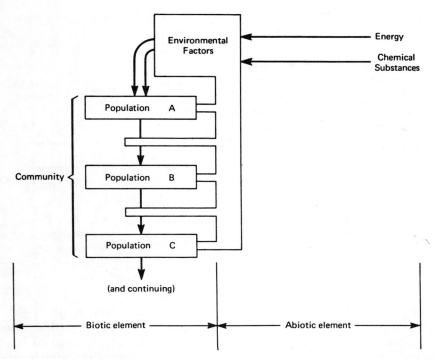

Fig. 1-1. Basic features of an ecosystem and the fundamental relationships within it. Ecosystems are composed of biotic and abiotic elements. The biotic element is formed from groups of similar and related organisms, the *populations*. These are linked together by functional interdependence to form *communities*. The abiotic element is involved in processes of energy absorbtion and release by the biotic element, and the uptake and release of chemical substances. Both these vital activities of an ecosystem are subject to environmental influences, as are the intracommunity and intercommunity biological interactions.

system: the total amount of organic material which these include represents its *biomass*. The biotic element can be categorized into *producer, consumer* and *reducer* components.

In the great majority of natural ecosystems where solar radiation is the initial energy source, producer organisms are invariably green plants. In terrestrial ecosystems these are for the most part the familiar flowering plants that constitute the dominant forms of vegetation over much of the modern land surface. In the oceans, the producer organisms are the microscopic free-floating *phytoplankton,* and more especially the group known as diatoms. All these producer organisms are *autotrophs,* utilizing the absorbed energy of sunlight and the composite pigment chlorophyll to synthesize sugars from carbon dioxide in the process known as *photosynthesis.* The *reducer* organisms of ecosystems make use of chemical

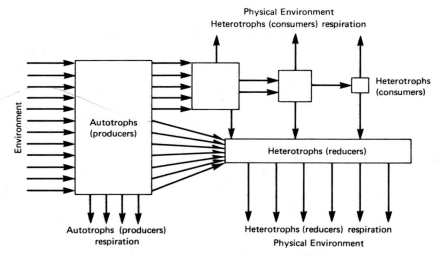

Fig. 1-2. Relationship between the basic ecosystem elements, illustrating the flow of energy and materials between the abiotic and biotic elements.

energy contained in these photosynthetic products and in the organic substances elaborated from them by the producer organisms. Reducers are therefore *heterotrophs.* When portions of the autotrophs such as dead leaves, or when whole organisms such as a windblown tree, fall to the ground, or die in a marine ecosystem, they are invaded by populations of reducer organisms (saprobes). These convert the organic compounds elaborated by the producers to inorganic nutrients in processes commonly described as *mineralization* and *nitrification.*

Although an ecosystem composed solely of producer and reducer organisms could be self-sustaining, few are encountered in the contemporary world which lack a second heterotrophic element in the form of *consumer* organisms, normally animals. This second heterotrophic element, like the reducers or decomposers, derives its energy from the elaborated organic materials synthesized by the producer element, and in its turn synthesizes further organic compounds. Those heterotrophic populations feeding directly upon plant populations are known as *primary consumers* or *herbivores.* Those which feed on the primary consumers are known as *secondary consumers* or *carnivores.* The carnivores are again divided into first, second, and even third-and fourth-level carnivores, the final level frequently being known as *top carnivores.* Animal parasites are a special type of carnivore.

The producer, reducer, and various categories of consumer populations are decribed as *trophic levels.* The broad functional relationships between these various trophic levels are considered later in this chapter.

More specific relationships such as those between populations in the same group and those between populations in different groups, interactions between these, and their association in *communities* are considered at greater length in subsequent chapters.

Energy Input and Utilization

For the great majority of natural ecosystems, whether terrestrial or aquatic, the primary source of energy is *solar radiation*. Following Odum, such ecosystems may be described as *autotrophic ecosystems,* distinguishable from *heterotrophic ecosystems* in which the energy input is from a source in the form of *chemical energy*. Most frequently heterotrophic or heterotroph-based ecosystems are portions of autotrophic or autotroph-based ones. One familiar example is provided by the benthic marine communities whose energy source is the chemical energy of the organic material sinking to the ocean depths. Another is illustrated by the various coprophilous communities which form microecosystems on fecal matter deposited by consumer organisms.

Not all autotrophic ecosystems necessarily derive their energy from solar radiation, for example the so-called *sulfur bacteria* are the producer element of an ecosystem which derives its energy from the decomposition of inorganic sulfur compounds. Iron bacteria are the autotrophs in a similar microecosystem deriving its source energy from the reduction of inorganic compounds. The absorption of energy into ecosystems whatever form this takes, and this transfer from one element to another, are considered again later in this chapter.

Nutrient Input

The existence of a heterotrophic reducer element in an ecosystem implies that nutrient cycling may take place, but this does not necessarily follow. The inorganic nutrients which are produced by mineralization of organic waste materials could be rendered unavailable to producer organisms, either by removal from the ecosystem by leaching or by incorporation in some unassimable chemical form. Some losses in this way are inevitable in most ecosystems. They are made good by a continuing nutrient input from such sources as the weathering of rocks, addition to the land and water surface of materials carried down by precipitation in the form of rain or snow, and by transfer from other ecosystems. In assessing such inputs and losses it is conventional for ecologists to prepare *nutrient budgets* (Fig. 1-3). These are considered in more detail later in this chapter.

When nutrient budgets are examined it becomes very apparent that

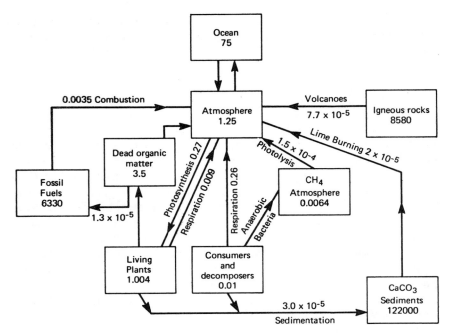

Fig. 1-3. A schematic nutrient budget for carbon—illustrating how it is possible to prepare estimates of the quantities of this essential element involved at each step of the carbon cycle. The question of transfer between available and nonavailable nutrient pools is considered in the next chapter. From the figures given here, which express the amount of carbon in each stage of the cycle in kilograms per square meter at any one moment of time, it will be noted that the use of carbon and production of carbon dioxide in industrial activity vastly exceeds the amount involved in the respiration of organisms. Moreover it is still increasing dramatically (see Fig. 1-16). (Based on data from Bowen, 1966).

of all the aspects of ecosystem structure and function, this is the one in which human interference with ecosystems is most obvious (Fig. 1-12). In our domestic ecosystems we constantly remove large quantities of nutrients in the form of agricultural, horticultural, and forestry produce. Sometimes we make good a portion of these losses by returning materials in the form of organic wastes or artificial fertilizers. At other times we merely exploit the ecosystem until it has been degraded to a level which no longer provides us with the same economic returns. We then abandon it or exploit it for some other purpose. In either case the effect of our removal of nutrients is radically to change both the structure and function of the ecosystem. Particular aspects of this exploitation are considered further in this chapter and the succeeding one, and are discussed in several of the volumes in this series.

Characteristics of Ecosystems

The *biosphere,* or global abiotic and biotic components of the world ecosystem, includes as has been seen the atmosphere and its oxygen, carbon dioxide, water vapor and other gases, and suspended particles. Together with the geological, chemical, and physical features of the totality of our habitats, these are sometimes grouped under the term *ecosphere* (Cole, 1958). An ecosystem can therefore feature the whole ecosphere and be as large as the earth. It can also be very small, as for example the aquatic ecosystem included within the clustered leaves of an epiphytic bromeliad in a tropical rain forest, or living forms which successively develop on a mouse pellet.

Ecosystems tend to become diversified and stable (Odum, 1969). The extent of information which they contain determines their diversity, the various feedback mechanisms which exist within them maintain their stability, and achieve a steady-state relationship between the component populations and communities and their environment (Figs. 1-4 and 1-5).

Ecosystems are not however static when stable; information is constantly being accumulated as they continue to adapt and evolve. This accumulating information permits further adjustment between the *biotic* component of ecosytems, the populations and communities of living organisms, and the *abiotic* element or environment. As we shall see shortly, they consequently exhibit the phenomenon of *ecological succession.*

Whereas populations can be argued to have at least sometimes an observable reality, and communities may fall within definable parameters, ecosystems are quite arbitrary as to range and size, and their definition is entirely a conceptual matter. They do not lie within any specific physical or chemical parameters. Sometimes however a physical barrier such as a river, a sea, or a mountain chain, a fence limiting grazing animals, or the edge of a wildfire burn may impose a physical limit in a particular direction.

The world is not therefore made up of a specific number of ecosystems, each with clearly definable borders. The number of ecosystems which we recognize and their size, is entirely a matter of convenience. We may chose to define an ecosystem which more or less coincides with what natural historians called a *biome* or *formation* (Fig. 1-4). This is a group of communities including particular plant, animal and microbial populations and characterized by specific forms of the most conspicuous plants, such as the deciduous broad-leafed tree species of the Northeastern

Fig. 1-4. The Boreal Forest provides an example of a large
ecosystem. It stretches across the Canadian Shield, forming
a broad belt of *taiga* between the tundra of the arctic
regions and the deciduous summer forest, which is effectively
the most northerly portion of the North Temperate Zone in
which extensive agriculture is possible. This ecosystem is
characterized by pines and spruce, with herbivores such
as caribou and carnivores like wolves.

forests or small-leaved evergreen shrubs in the case of the chaparral of
the Pacific Southwest. Alternatively, the selection of a somewhat smaller
set of parameters for an ecosystem might align it with what has been
formally called an *association*. For example the oak-pine *association* of
the Northeastern forests of the United States. Or an ecosystem could be
limited to some smaller section of such a grouping, like a river bank
(riparian) community and its environment. Also definable within ecosystems are the man-made assemblages resulting from the degradation of

Fig. 1-5. An example of a microecosystem—an epiphytic growth of Stag's horn fern (*Platycerium*) on a forest tree in Mozambique. Each fern plant is a miniature ecosystem in itself, receiving water and minerals from outside the system, supporting a saprobial microflora and a microfauna within the basal fronds which slowly build up a humus mat on the trunk of the tree, and form a number of heterotrophic food chains.

natural communities, or the entirely synthetic living systems which man can construct for one purpose or another (Fig. 1-11).

Further Structural Features of Ecosystems

The laws of thermodynamics require that when energy is transferred between the various trophic levels in an ecosystem there is a degree of loss due to the conversion of some of this energy to heat energy, which

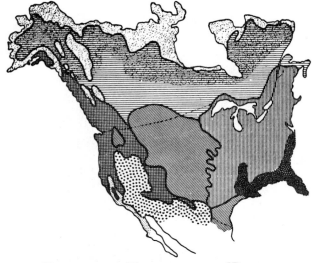

▦ Tundra ▦ Boreal Forest ▦ Deciduous Forest ▦ Montane, Subalpine, Coastal Forests
▦ Grassland ▦ Southern Conifer Forest ▦ Desert ▦ Chaparral

Fig. 1-6. Major biomes of North America. A biome scheme such as this has much convenience in terms of field ecology, but has a lesser significance if its basis is critically examined in terms of theoretical ecology, when many of the apparent boundaries between biomes will tend to disappear. Nevertheless we can equate these biomes with ecosystems, as has been done for example with the chaparral biome or ecosystem illustrated in Fig. 2-15. Certain of these biomes have also been recognized for the purposes of the International Biological Program.

is then dispersed into the environment. As the consumer organisms are generally animals, and frequently have extensive locomotor activities, considerable loss of energy also occurs from this source. As a rule, the amount of energy which is available to one trophic level from its predecessor amounts to approximately 10 percent of the energy it receives (Kozlovsky, 1968). In a given ecosystem there is therefore what is described as an *energy pyramid* which soon reaches a practical limit beyond which there is insufficient energy for the support of further trophic levels (Fig. 1-7). This forms what is commonly known as an *Eltonian pyramid,* after a pioneer British ecologist Charles Elton, who first devised this schematic method of expressing basic ecosystem structure. In terrestrial ecosystems there are rarely more than two to four levels of secondary consumers; marine ecosystems may have one or even two more than this. In addition to their use for the schematic representation of energy flow, Eltonian pyramids may also be used to indicate matter (*biomass*) and individual numbers of organisms within an ecosystem (Fig. 1-8).

This study of the process of energy transfer between trophic levels is

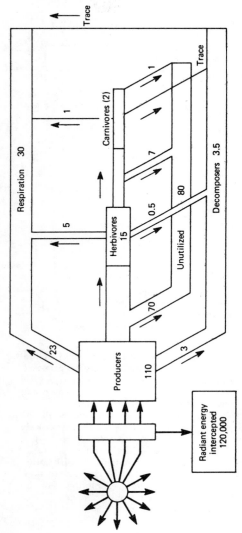

Fig. 1-7. An energy pyramid, expressing the directions of energy flow and exchange between the basic components of a hypothetical ecosystem. The figures shown represent energy values in calories per square centimeter per year. The scheme shown the unidirectional flow of energy through the ecosystem. The energy which enters as solar radiation can be expended in three basic ways—in respiration, in transfer to an unutilized pool, and in biomass production diminished appreciably at each trophic level: the amounts of energy involved in this instance are often isolated and shown in pyramic form (Fig. 1-8).

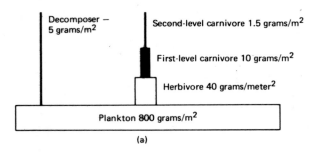

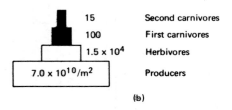

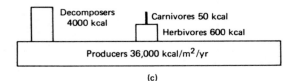

Fig. 1-8. Eltonian pyramids. Schematic expressions of biomass (a) and numbers of individuals (b) may be prepared
in the same manner as for energy distribution (c). They
usually show the same pyramidal decreases progressing
from early to later trophic levels in an ecosystem.

As would be anticipated, in the marine ecosystem
illustrated (a) the biomass of carnivores, and particularly
that of the top carnivores, is immensely less than that of
the producer organisms. Likewise the numbers of carnivores are very much lower in comparison with other
trophic levels. Carnivores also tend to be larger than their
prey, which still further restricts the numbers of individuals encountered in a given trophic biomass. (Data
from various sources.)

sometimes known as *bioenergetics* (Phillipson, 1966). The most promi-
nent populations in these communities, the ones through which most
energy transfer in the ecosystem takes place, are known as *dominants*.
Populations connected by a successive series of consumer-consumed relationships are said to form a *food chain* (Table 1-1 and Fig. 1-9). Under
natural conditions food-chain relationships multiply and diversify to

form a complex known as a *foodweb*: The functional relationships within communities originate in such interwoven *foodwebs*. The relationships between the populations involved in this energy transfer are expressed schematically by constructing diagrams of such *foodwebs*. In Fig. 1-10 the populations located at each trophic level form such an identifiable foodweb.

Foodwebs

The investigation of foodwebs in an ecosystem and attempts to break these down further into their interlocking *food chains* provides information as to the arrangement of populations in consumer-consumed relationships. A food chain is hypothecated by successively identifying any populations which use for food a particular population in the trophic level immediately below their own, working through an ecosystem from producer to final consumer levels. The relationship can be experimentally confirmed by direct observation, by examining stomach contents, or using radioactive tracers. Examples of particular food chains which might feature in the foodwebs of common terrestrial and marine ecosystem are listed in Table 1-1.

TABLE 1-1

Examples of simple food chains. As is frequently the case, the marine food chain contains more steps than the terrestrial example. One explanation advanced for this common difference is that marine systems are older, and have had a longer period in which to evolve.

Tropic Level	Marine Food Chain	Terrestrial Food Chain
Producers	phytoplankton	grasses
Primary consumer (herbivore)	herbivorous zooplankton	cotton-tail
Secondary consumers (1) (general carnivore)	anchovy	fox
Secondary consumers (2) (intermediate carnivores)	herring	
Secondary consumers (3) (top carnivores)	tuna	

Synthetic and Exploited Ecosystems

Under natural conditions food chains in reality are rarely so elemental as expressed schematically in such diagrams. Relationships as simple as this are not found in natural foodwebs (Fig. 1-10) but are generally encountered only in synthetic ecosystems. These are either devised in laboratories for experimental purposes, or those maintained in cultivation for agricultural, horticultural or forestry purposes (Fig. 1-11).

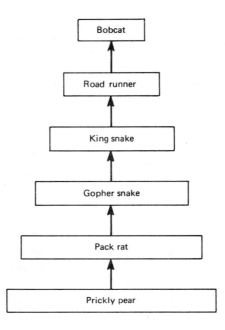

Fig. 1-9. One food chain from chaparral in Southern California. The consumer-consumed relationship between these dominant organisms in this "predator chain" can be isolated from the generalized interdependence expressed as a foodweb in Fig. 1-10. It would be necessary to do so however only if energy exchange or population dynamics were being investigated experimentally in laboratory studies. In such "predator chains" the consumer at each successive trophic level is generally larger than the consumed animal. Commonly the order of increase is twice the mass and 25 percent of the linear dimensions. Such increases strictly limit the number of successive trophic levels in a food chain, as illustrated here.

Each box is not proportional to the number of animals present at each stage, for as illustrated in Fig 1-8 (b) this would involve a drastic reduction. As discussed in the text, energy considerations still further limit the population densities of successive consumers in a food chain.

For instance the corn-pig-man food chain represents a typical and simple example of the monoculture for which modern agriculture commonly strives. Even this artificial ecosystem is however far more complex

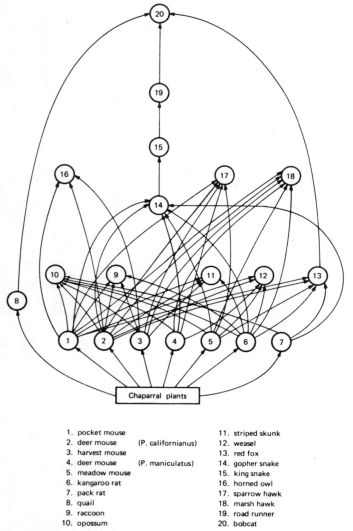

1. pocket mouse
2. deer mouse   (P. californianus)
3. harvest mouse
4. deer mouse   (P. maniculatus)
5. meadow mouse
6. kangaroo rat
7. pack rat
8. quail
9. raccoon
10. opossum
11. striped skunk
12. weasel
13. red fox
14. gopher snake
15. king snake
16. horned owl
17. sparrow hawk
18. marsh hawk
19. road runner
20. bobcat

Fig. 1-10. The basic foodweb of the chaparral ecosystem, as illustrated in Fig. 2-15. Deliberately omitted from this is a shrub-mule deer-cougar food chain, which is now sometimes absent because of human disturbance of this ecosystem resulting from the hunting of both animals. Earlier hunters from middle Pleistocene times are believed to have disrupted yet other food chains of this ecosystem by exterminating such animals as sloths, camels and horses and thus causing their predators such as "lion" and saber-toothed cat also to pass to extinction.

The chaparral ecosystem we have today, even in areas where we refer to it as "undisturbed," has been vastly and irreversibly changed by approximately 20,000 years of human occupation.

Omitted from this diagram entirely are insects. Every animal in this foodweb probably eats one or more insect species at some stage.

than is indicated by the scheme. The reducer trophic level is composed of many series of reducer food chains of saprobe populations forming quite complex foodwebs which are not represented in the diagram.

Human occupation of particular natural ecosystems has extremely important consequences upon the populations of their food chains and foodwebs. The extent of human interference with the structure of a natural ecosystem, and therefore with the measure of diversity which it is possible to conserve in the ecosystem, will be dependent on the degree to which these food chains and foodwebs are disturbed (Fig. 1-12).

Basic Interaction Within Ecosystems

It would be entirely incorrect to regard these biotic relationships in the food chains and foodwebs within an ecosystem as static. There are two major types of interface in the ecosystem at which species populations interact dynamically (Harper, 1961). The first kind of interaction is between populations at the same trophic level. This results in *competition* between species and is further discussed in Chapter 5, which deals with population interaction. Competition can result in the extinction of the less "fit" species. Thus loss of diversity in natural ecosystems may occur when we insert our own species as a vegetarian to *compete* with other herbivore populations at the primary consumer trophic level. For example, we drive off the crows feeding on the corn. As a meat eater we can compete with other predatory populations in the top carnivore trophic level, killing the wolves, cougars, and coyotes which predate deer. When we compete with reducer populations by removing accumulated organic materials or *biomass* from the ecosystem, as when we mow hay, we deprive these decomposer populations of essential energy sources and nutrients.

The other interface within the ecosystem at which biological components interact is between consumer and consumed populations. These may be plant-herbivore, herbivore-carnivore or general carnivore-specialized carnivore interactions. The feedback mechanisms which evolved in such relationships in natural ecosystems usually produce a steady-state relationship at this interface. Unlike the competitive interaction *within* a trophic level this relationship ensures the *survival* of all the species involved, rather than the *extinction* of the less "fit." The disruptive influence of human populations on this second kind of interface in natural ecosystems results from overriding such feedback mechanisms, with a consequent overexploitation of some food resources (Fig. 1-12). The practical consequences of this overexploitation are treated further in Chapter 5 and in a number of the volumes in this series.

Other important considerations besides these evolutionary effects

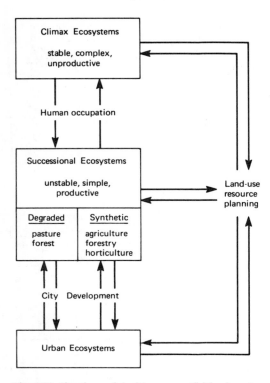

Fig. 1-11. Simple model of human activities in relation to ecosystem development. Stable mature ecosystems of a "climax" type are degraded to less stable ecosystems as a result of human interference with their structure and function. These are still further degraded during the creation of cities. Resource-utilization planning can balance this system of climax-successional-urban ecosystems at a particular level and hold it in a steady state.

may be developed from such trophic interactions, more especially those relating to the process of energy absorption and transfer (Turner, 1968). The amount of energy involved in these particular functions of ecosystems is expressed in a quantitative relationship which is known as *productivity*.

Productivity

Just as the concept of trophic levels defines the primary structure of an ecosystem, so *productivity* expresses its functional activities. Properly speaking, productivity is a *rate,* the amount of energy, commonly meas-

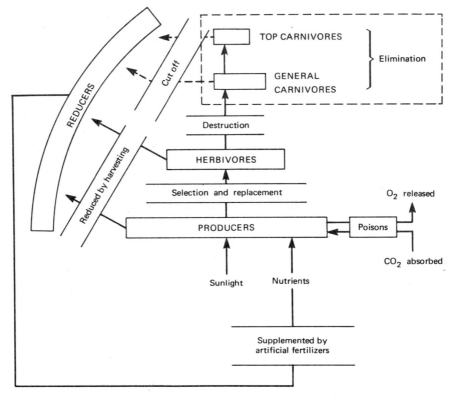

Fig. 1-12. Human disturbance of natural ecosystems—has many sources, the commonest of which are illustrated schematically here. Catastrophic disturbance of population regulation mechanisms can result from the killing of predators as "vermin," or the selective elimination of herbivores and the introduction of others. The dependent heterotrophic microecosystems may be disrupted by reduction in the quantity of biomass previously available to them through cropping and harvesting. Correction of this kind of interference by addition of nutrients may radically disturb the producer trophic level. Dependent cycles such as the self-cleansing and self-replenishing system which maintains the global carbon dioxide/oxygen ratio may be disrupted by release of poisons into the air or water.

ured in terms of units of heat energy (calories), produced in a unit area (usually one meter square) per unit time (one year). This rate is known as the *gross productivity,* and in natural ecosystems it results from the photosynthetic activities of the green plants which are the sole organisms in the producer trophic level. Gross productivity is therefore a finite rate, limited both by the number and activity of the producer organisms and by the amount of solar energy available (Ryther, 1963; Westlake, 1963; Engleman, 1966; Whittaker, 1966; Whittaker and Woodwell, 1968; Whittaker, 1970).

If productivity is regarded as an *amount* rather than a *rate* it is more usually referred to as *production,* and it can be known as *biomass production.* The biomass or standing crop as already noted is the total amount of organic matter produced within the ecosystem. In agricultural ecosystems or in forestry production it is known as the *yield,* and is expressed in bushels per acre or by some similar measure. This is not the total biomass or production, but only the portion removed from the ecosystem.

Primary Productivity

The producer organisms themselves will utilize some of the photosynthates in their own respiratory processes, so that the gross productivity is reduced by the extent of this respiratory activity. When this respiration value is deducted, the rate which remains is described as the rate of *net productivity.* In fact this is the *net primary productivity,* because it ignores the amounts of energy utilized by the consumer and decomposer organisms of the ecosystem. The amounts of energy available to these other two trophic levels are usually described as *secondary productivity.*

Secondary Productivity

Secondary productivity represents the rate of energy transfer between *consumer* trophic levels, and also between *decomposer* trophic levels. It has to be a whole order of magnitude less than gross primary productivity, because as can be noted in the energy budget in Fig. 1-8 and elsewhere, only approximately one-tenth of the amount of energy absorbed by one trophic level becomes available to the next trophic level. For many ecological purposes, secondary productivity can therefore be ignored without introducing great error, especially having regard to the many inaccuracies involved in productivity measurement. In animal husbandry, and to pastoral and hunting human populations however, it is of vital concern.

In these situations, human populations may be able to harvest a considerable proportion of the secondary production, as in the corn-pig-man ecosystem already discussed. It is not ever possible to harvest the whole, for reducer organisms remove that portion of the energy absorbed by primary or secondary consumers which is converted to chemical energy in excretory and biomass waste.

Too little information has as yet become available on secondary productivity to permit an adequate analysis such as is provided later in this chapter for primary productivity. What there is suggests some anomalies—for example, that an ant colony which appears efficiently organized, is very inefficient in terms of the amount of secondary productivity. One of the primary objectives of the several biome studies included in the International Biological Program is the measurement of secondary productivity. The urgent need for such information is emphasized in many of the volumes in this series.

Methods of Measuring Productivity

The ideal method of recording gross productivity in an ecosystem would be by measuring energy flow between trophic levels. However, no direct method of doing so has yet been evolved, and indirect methods have to be used. The procedures which are followed have been well described by Odum (1959) and others (Steeman, 1963). They may be deduced logically from the basic equation for photosynthesis which can be expressed as

$$Energy + 6CO_2 + 6H_2O = C_6H_{12}O_6 + 6O_2$$

This is the familiar way of expressing the reduction of carbon dioxide to produce a hexose sugar and molecular oxygen.

When converted to its calorific and element values, this basic equation may be expressed as

$$1,300,000 \text{ cal radiant energy} + 106CO_2 + 90H_2O$$
$$= 13,000 \text{ cal potential energy in } 3258 \text{ g protoplasm} + 154O_2$$
$$+ 1,287,000 \text{ cal heat dispersed}$$

Gross productivity can be indirectly estimated experimentally by assessing any of the quantities in this last expression. These procedures are, in order starting from the left-hand side, measuring the amount of energy received, the amount of carbon dioxide absorbed, the weight of resulting protoplasm, and the amount of oxygen released. Generally wherever possible more than one of these procedures is followed in order to provide a check on the experimental results.

The Chlorophyll Method

The chlorophyll method comes the nearest to an estimation of the quantity of energy utilized in an ecosystem and therefore its *gross productivity*. It depends on the observation made by F. Gessner in 1949 that the total quantity of the pigment in a specific area is characteristic of an ecosystem and remains relatively constant under constant conditions. The amounts of chlorophyll present in samples obtained at particular intervals of time and estimated by spectrophotometric analysis therefore provides estimates of variation in gross productivity.

Some estimates of productivity obtained in this way are shown in Table 1-2. The main objection to this experimental procedure, a very

TABLE 1-2

Seasonal variation in primary productivity as determined by the chlorophyll method. Figures compare the grams of carbon dioxide fixed per square meter per day as between an inshore station up to 50 m. depth (A) and an offshore station up to 1000 m. depth (B) off New York.

	Sept.	Dec.	Feb.	March	April	June
A	0.2	0.5	0.6	0.7	0.9	0.3
B	0.1	0.1	0.1	0.5	0.1	0.2

Data extracted from Ryther and Yentsch, 1958.

substantial one, resides in the fact that the relative amount of chlorophyll is not the only factor that may influence the rate of photosynthesis in producer organisms.

Carbon Dioxide Absorption

The commonest method of measuring productivity utilizes an instrument known as an infrared gas analyzer, which records the amount of carbon dioxide in air by spectrometric determination of its infrared light absorption. This provides a value for the *net productivity*. The gross productivity is obtained by running the same experiment in the dark, and measuring the amount of carbon dioxide released in respiration. Added to the original figures for net productivity, this provides a value for gross productivity. Some measurements of gross productivity arrived at by this method are provided in Table 1-3.

Harvest Method

Another very common method of assessing gross productivity requiring less sophisticated equipment is termed the *harvest method*. It consists simply of weighing and chemically analyzing all producer biomass, leaves, stems, roots, flowers and fruit. Generally speaking, the chlorophyll estimation technique is applied in marine situations, the infrared gas analyzer in autecological studies concerning small plants and animals, and the harvest method in synecological investigations.

The harvest method has several disadvantages. Like the carbon dioxide absorption method, it is actually a measure of *net* productivity. The gross productivity can only be estimated by the addition of a calculated value for respiration which has been independently obtained. Another difficulty is that a portion of the producer biomass in a given ecosystem may be removed by herbivores before it can be measured. If the herbivores are insects, it is possible to limit their activities by treatment with insecticides while readings are being taken. If they are by contrast large grazing animals as in agricultural studies on domestic animals, it is possible to exclude them from the area while observations are made. In marine ecosystems, however, the difficulties inherent in this method for such reasons will be apparent.

Examples of gross productivities estimated using the harvest method are provided in Tables 1-3 and 1-4. For comparison, the gross productivity of a natural ecosystem can be contrasted with figures obtained by agronomists for an agricultural ecosystem. In agricultural investigations it is customary to express productivity not in terms of gross productivity, but as a value of net secondary productivity expressed as the number of domestic animals the ecosystem will support. This takes such forms as one cow per 30 acres, or one sheep per 10 acres, referring to the net sec-

TABLE 1-3

Productivity as assessed by the harvest method—the flow of energy through a plantation of Scots pine (*Pinus sylvestris*). Figures express the value of production calculated as units of 10,000,000 kilocalories per hectare.

	Total Net Production, 180		
Tree production, 105	Litter production, 74		Ground flora production less than 1
Living trees, 61	Lumbered trees, 44	Decomposed, 68	Accumulated, 6
.	Removed, 31 — Left in roots, 13		

(Based on data from Ovington, 1962.)

ondary productivity of a grazing area. Many of the productivity statistics reproduced in the volume in this series dealing with food resources are based on the harvest method.

Oxygen Production

The last major procedure for the measurement of gross productivity, the measurement of oxygen, is also more applicable to aquatic ecosystems. It is unsuitable for terrestrial situations because in them both the con-

TABLE 1-4

Productivity in an aquatic ecosystem as recorded by the harvest method—in this particular instance net secondary productivity. The harvest method is just as applicable to the measurement of net secondary productivity as it is to the determination of net primary productivity. These figures illustrate the effect on productivity of adding nutrients and additional food into an aquatic ecosystem. They also show the reduction in total net secondary productivity in an aquatic ecosystem when a carnivorous fish trophic level is present.

Trophodynamic Status of Community	Net Secondary Productivity, g/m² per year
Unfertilized waters	
North Sea, herbivore-carnivore	1.68
Great Lakes, herbivore-carnivore	0.9 - 0.8
Fish ponds, U.S. (with carnivores)	0.21- 18.1
Fish ponds, Germany (without carnivores)	11.2 - 30.0
Fertilized waters	
Fish ponds, U.S. (with carnivores)	22.4 - 56.0
Fish ponds, Germany (without carnivores)	99.7 -157.0
Fish ponds, Phillipines (without carnivores)	50.4 -100.0
Fertilized waters plus added food	
Fish ponds, Hong Kong (without carnivores)	224.0 -448.0

Data from various sources.

sumer and reducer elements of the ecosystem tend immediately to absorb oxygen released by producers. The technique is usually known as the "light and dark bottle" method. Two bottles, one clear and the other darkened, are filled with a suspension of the producer organisms and left in an appropriate situation. After a given interval the oxygen content of the bottles is estimated, and from this the amount of respiration which has occurred in the dark bottle is calculated. This is added to the net productivity obtained from the value of oxygen produced in the light bottle, thus giving a figure for the gross productivity. Typical results obtained form the light and dark bottle method are provided in Table 1-5.

Radioactive Isotopes

Inaccuracies in the light and dark bottle method arise from the growth of bacteria on the bottles. Their metabolic activities disturb the oxygen-carbon dioxide values recorded. Phytoplankton productivity is now most commonly estimated by using labeled carbon. This is done by introducing sodium bicarbonate incorporating the radio-nuclide carbon-14 into the bottles. The amount of C^{14} taken up is a measure of the gross productivity of the phytoplankton after one or two hours, and of net productivity after one or two days (Table 1-6).

TABLE 1-5

Gross primary productivity as determined by the light and dark bottle method in various aquatic ecosystems. These figures illustrate both the high productivity of coral reefs and estuaries, and the extremely high productivity of polluted aquatic ecosystems.

Ecosystem	Gross Productivity, g/m^2 per day
Sargasso Sea	0.5
Oligotrophic lake, Wisconsin	0.7
Long Island Sound	3.2
Fertilized pond, North Carolina	5.0
Lake Erie (polluted, summer)	9.0
Pacific coral reef	18.2
Estuary, Texas	23
Polluted stream, Indiana	57

Measurements of Secondary Productivity

The harvest method as already indicated may be applied to the calculation of secondary production, that is the amount of energy which is absorbed and released by the herbivores and carnivores of the consumer trophic levels and the saprobes of the decomposer level. In terms of the consumer trophic levels, secondary productivity is equivalent to the respiration of the animals concerned, to their heat loss and energy utilization in movement, together with the weight of dead animals and their excretory products.

TABLE 1-6

Primary productivity as determined by the uptake of C-14. Figures represent the relative uptake of C-14 in surface sea water enriched as indicated, expressed as a ratio to that fixed by unenriched controls. Data modified from Menzel and Ryther, 1961, showed that addition of metals (actually the addition of iron) doubled the productivity.

Type of Enrichment	Ration of Increase in Productivity
Nitrate and phosphate	0.9
Nitrate, phosphate and metals	2.1
Metals only	2.0
Vitamins only	1.0
Complete enrichment	1.8

More especially under agricultural conditions in synthetic or degraded ecosystems, secondary productivity must be carefully related to seasonal conditions. Thus in a country which practices intensive cultivation such as Japan, the secondary productivity of agricultural ecosystems is maintained at a high level continuously. In some parts of America, where agricultural operations are undertaken only during a favorable growing season, secondary production may fall to very low values during the inclement period.

Levels of Productivity

The level of productivity of a particular ecosystem is obviously of vital concern in assessing its potential for human food production. The four broad categories of productivity commonly but arbitrarily recognized are as follows:

1. *Highly productive ecosystems.* These have a gross productivity of five to 20 grams per square meter (g/m^2) per day, and a net production which can exceed 3,000 g/m^2 per year. Such highly productive communities include areas like river estuaries and their dependent salt marshes, coral reefs, flood-plain forest, coastal redwoods and some agricultural ecosystems such as rice paddies and sugar cane plantations.

The abiotic elements of such ecosystems show a number of features in common. They exist in warm-temperate regions and they have an abundant water supply. Usually their nutrient levels are constantly maintained by immersion in a continuously moving nutrient solution as in salt marshes, or by periodic inundations as in sugar plantatons, rice paddies, coastal redwoods, estuaries, and flood-plain forests. The key to high productivity in all these instances is therefore a continuous supply of available nutrients associated with an absence of limiting moisture and temperature levels.

2. *Average productivity ecosystems.* These have a gross productivity of three to five g/m^2 per day and a net productivity of 1,000 to 2,000 g/m^2 per year. They include most temperate forests and agricultural crops, shallow lakes and the off-shore waters over the continental shelves. Again, water supplies and temperatures are not limiting, at least not all the time, and nutrients are available in fair supply.

3. *Less productive ecosystems.* These have a gross productivity of $1/2$–3 g/m^2 per day and a net productivity of 2 to 1,000 g/m^2 per year. They include most grassland ecosystems and cereal crops, many shrublands such as the Pacific Coast chaparral, open woodlands and savannas, oligotrophic lakes and other open waters. The restricted productivity of such ecosystems may result from a limited supply of water at some time of year, or it may arise because available nutrients are at a much lower level than in the previous categories.

4. *Ecosystems of low productivity.* These have a gross productivity up to $1/2$ g/m^2 per day, and a net productivity of about 200 g/m^2 per year. They include the tundra of high latitudes or high altitudes, deserts, some open seas and lakes.

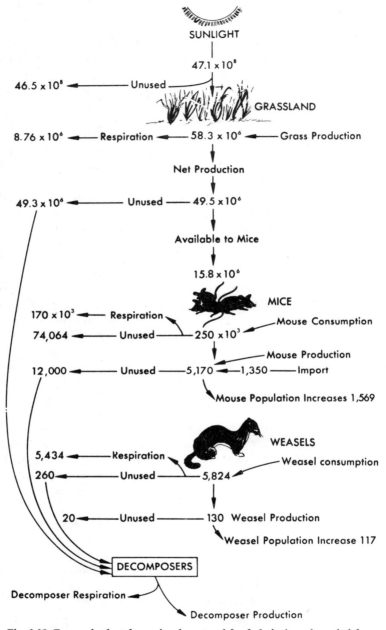

Fig. 1-13. Energy budget for a simple natural food chain in a degraded forest ecosystem in southern Michigan. The energy flow between the plant-meadow mouse-weasel, producer-herbivore-carnivore trophic levels is typically limited by the energy lost in other ways as shown. Figures express energy values in calories/m²/year. (After F. B. Golley, in *Ecological Monographs* 30 (2): 187-206, 1960; from A. S. Boughey. *Ecology of Populations,* Macmillan, 1968, reproduced with permission of the publishers.)

The operation of certain abiotic factors which may be limiting productivity in these ecosystems such as temperature, or available water, is considered further in the next chapter, as is the question of the effect of available nutrient levels. The more general influences of abiotic factors on these and other ecosystem productivity are considered by preparing energy and nutrient budgets.

Energy Budgets

It is not easy to present energy-flow budgets for natural communities because, as has been seen, the foodwebs of which they are composed are exceedingly complicated. However, by selecting a relatively simple example, and by ignoring all populations but the dominant ones, it is possible to arrive at an evaluation such as is expressed in Fig. 1-13. This is the energy-flow budget for a plant-meadow-mouse-weasel food chain in an old field community of a degraded forest ecosystem in southern Michigan (Golley, 1960). It well illustrates the limitations on energy transfer between trophic levels.

In this example about one percent of the total solar energy in the incident radiation is converted into plant tissue. Of this plant biomass, the meadow mice, which are the dominant herbivores of this ecosystem, consume only 2 percent of the available energy. Some 10 to 20 percent of this potential energy in the consumer trophic level is removed by herbivorous insects. At the secondary consumer or carnivore level the weasels, which depend almost exclusively on the meadow mice for food, utilize only 30 percent of this mouse biomass. Looking at the energy actually absorbed by each trophic level of this simple ecosystem, the consumers respire 15 percent, the mice 68 percent, and the weasels 93 percent. At the carnivorous trophic level therefore the gross productivity or the total amount of energy available is almost entirely utilized in respiration and locomotory activity. So little energy remains that, allowing for the comparatively small amount entering this ecosystem, it is impossible to support a further carnivore level to prey upon the weasels.

It will also be apparent from this energy budget, that a greater amount of energy is available to the herbivorous mice than to the carnivorous weasels. When they insert themselves into carnivorous trophic levels, human populations similarly limit, usually by a whole order of magnitude, the amount of energy which they could derive from the same ecosystem by adopting a vegetarian diet and remaining in the herbivore trophic level.

In any agricultural ecosystem such as the corn-pig-man system referred to previously, nutrients are removed from the system, in this instance in the form of pork, and are rarely returned directly because human excretory products are disposed elsewhere, as are other human wastes such as corpses. The nutrients removed have to be replaced by adding artificial fertilizers to the system, or some form of "natural" manure. Such additions are necessary if productivity is to be prevented from dropping appreciably. In order to examine productivity further it is necessary therefore to consider the matter of biogeochemical cycling.

Biogeochemical Cycling

The living matter of the ecosystem, its biomass, is formed by *producer* populations, utilized and increased by *consumer* populations, supplemented and broken down again to its basic inorganic nutrients by *decomposer* or *reducer* populations. This cycling of nutrient materials through these different community levels is known, as already noted, as *biogeochemical cycling,* referring to its "earth chemistry" (geochemistry) origins. It is as vital to the functioning of the system as is the process of energy transfer, from which it is inseparable.

Nutrient Cycling

Along somewhat similar lines to energy budgets, nutrient budgets may be prepared to chart the cycling of particular nutrients through an ecosystem. Of the many nutrients essential to growth six elements are universal and in highest demand, oxygen, carbon, hydrogen, nitrogen, phosphorus and potassium. The first three are more or less freely available in atmospheric oxygen and carbon dioxide (Fig. 1-3), and surface water.

In a man-made ecosystem such as a farming rotation, nutrient budgets therefore especially concentrate on supplies of the second three. One of the most common agricultural experiments undertaken combines artificial fertilizers containing nitrogen, phosphorus, and potash, so-called NPK trials. Coupled with the use of multifactorial statistical analytical techniques, these provide information as to the levels of the three essential nutrients which are required in a particular agricultural ecosystem to achieve specific yields (Table 1-7). The design of such experiments and the analysis of the results has now reached a degree of sophistication that far exceeds similar investigations of natural ecosystems. The results of these agricultural experiments are potentially capable of a much more extensive application to natural ecosystems than has so far been generally attempted.

TABLE 1-7

The effect of major macronutrients on production, illustrated from a standard agricultural manurial trial. The effect of the application of nitrogenous, potassium and phosphate fertilizers on cotton yield was determined in a replicated plot experiment designed to permit multifactorial analysis. The data in the table summarize the results.

Treatment	Yield of Seed Cotton, Pounds per Acre
Control (no fertilizer applied)	187
N only	267
N, P	274
N, K	279
N, P, K	285

N applied as 150 lb per acre of ammonium sulphate.
P applied as 600 lb per acre of superphosphate.
K applied as 300 lb per acre of potassium sulphate.

In many natural ecosystems essential nutrients are obtained in less direct and more complicated fashion. The source of most of the nutrients is ultimately the rock crust of the earth, the study of whose chemical composition is known as *geochemistry*, hence the term *biogeochemical cycling* as already noted. Specific examples of the cycling of individual essential elements are considered in the next chapter.

Biogeochemical cycles in ecosystems involve transfer patterns between populations at the various trophic levels and a transfer rate between available and nonavailable nutrient pools. These are also inevitably associated with some losses, if only from leaching in drainage waters. In an agricultural system where some portion of the biomass produced is *harvested* the nutrients thereby removed are replaced by addition of artificial fertilizers. In natural communities, wind and rain carry in much of the needed nutrient replacements. *In situ* weathering of bed rocks may provide some or all of the rest.

Salt spray blowing off breaking waves has an obvious nutrient content in coastal situations. A small deciduous forest formed on the sand bar of Fire Island in Long Island Sound obtains approximately 95 percent of its required nutrients from this source. What is not so obvious is that these spray droplets can evaporate and their salt content be carried far inland by onshore winds. Air currents also transport nutrients in dust which has been blown from soil surfaces. Together these two sources may provide significant amounts of nutrients which are carried down again to the soil surface by precipitation in the form of rain or snow. The amounts of nutrient involved are indicated in Table 2-3.

Not only are nutrients recycled through ecosystems in this way, but as the volumes in this series on pollution and on pesticides describe, vaporized toxic substances such as pesticides and radionuclides may be carried in the atmosphere on dust particles and be brought down by precipitation into various ecosystems.

Nutrient Levels

While many other factors of the abiotic element of an ecosystem such as water supply, or temperature, can limit productivity as in deserts in the first instance or polar regions in the second, in most ecosystems its level is determined by the rate at which inorganic nutrients are re-cycled by reducers for renewed uptake by producer populations. If the end result of the activities of the reducer populations provides nutrients which are not immediately available to the producer trophic level, then productivity rates must fall. Odum presents the idea of regarding each geochemical element as existing in two *nutrient pools,* one located in the biotic element, the other in the abiotic element. The first is regarded as unavailable, the second as available, and between the two pools a steady-state balance is achieved whose rate of exchange determines the level of productivity of the ecosystem. As productivity is a *rate,* it is the amount of this exchange, the *transfer rate,* between the two pools which is significant, rather than the actual *total* of nutrients available. This total cannot however be entirely ignored, because the *absolute* amounts of nutrients in the unavailable pool are of significance when considering resources depletion.

The various geochemical elements which are essential to the proper functioning of an ecosystem are conveniently classified into macronutrients and micronutrients or trace elements. The nature of these and of nutrient pools is considered further in the next chapter where particular cycles such as the nitrogen cycle, the oxygen cycle, the hydrological cycle and the calcium cycle are discussed in greater detail. Many of our most serious environmental confrontations have resulted from deliberate or unwitting exploitation of the resource pool of available nutrients, modification of transfer rates, and disturbance of ecosystem nutrient transfer patterns. It is unhappily all too apparent that even at the present size of our world population, the amounts of available nutrients are far below what is required to maintain the whole of our global populations at the levels of affluence and sophistication which industrial technologies have now permitted in a number of the advanced industrialized societies (Committee on Resources, 1969).

As several volumes in this series describe, the preparation of nutrient budgets and the determination of productivity rates, now indicate the maximum limits of the human populations which we can support without causing degradation and overexploitation of particular ecosystems.

These are parameters which in a more restricted context have been investigated by agricultural economists. As already mentioned, more comprehensive investigations are now in progress in connection with the International Biological Program (IBP) which is designed to obtain information on the productivity of the basic biomes into which the world for convenience can be divided. These include grassland, desert, deciduous forest, coniferous forest, tundra, and tropical rain-forest ecosystems.

Nutrient Budgets

As with energy budgets it is possible to provide a nutrient budget for particular ecosystems (Chapman, 1967). The development of radioactive tracer technology has greatly facilitated such studies, and the diagram in Fig. 1-14 illustrates the kind of scheme it is possible to construct from the results of such investigations. These are more fully described in the

TABLE 1-8

A nutrient budget for a plantation of Scots pine (*Pinus sylvestris*), showing the distribution in this microecosystem of the macronutrients, nitrogen, phosphate, and potassium salts. The main portion of all three nutrients is contained in the pines and their litter. Logging, with subsequent decomposition of the litter will therefore result in a drastic reduction in the nutrient status of the ecosystem.

	Nitrogen	Phosphate	Potassium
Trees			
Total uptake	4,817	413	1,933
In living trees	453	41	150
Removed in lumber	161	14	98
Remaining in logged crowns and roots	704	75	279
Litter fall	3,499	283	1,406
Transfer from trees to soil	4,203	358	1,685
Ground flora			
Total uptake	2,058	182	876
Present in ground plants	40	3	40
Change in ground plants	−73	−5	−73
Litter fall from ground plants	2,131	187	877
Litter			
Total uptake (trees and ground plants)	6,334	545	2,562
In litter layers	409	28	34
Change in litter layers	+250	+19	+24
Released in decomposition	6,084	526	2,538
Average annual uptake by trees and ground plants	125	11	51
Average annual change in ecosystem	+11	+1	+3
Average annual removal in tree trunks	3	0.3	2
Average annual release by decomposition	111	10	46

(Based on data from Ovington, 1962.)

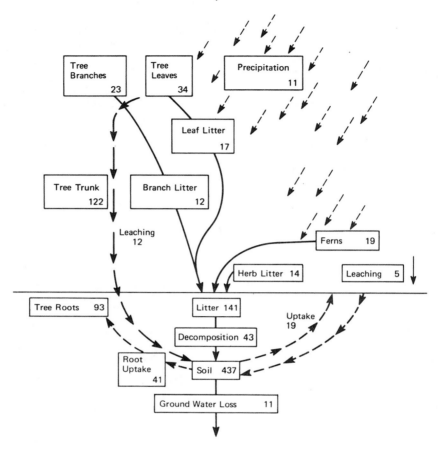

Fig. 1-14. Nutrient budgets—are prepared for a given ecosystem by a systems analysis of the processes of input, transfer and output of a particular chemical element, as shown schematically in this diagram of a hypothetical example of calcium distribution. The quantities of nutrient involved at each step then have to be determined experimentally, and can be entered at the appropriate points indicated by the boxes. (Based on figures extracted from Ovington, 1962).

next chapter. Quantitative information on nutrient flow in ecosystems is most usefully obtained by the study of whole catchment areas in an ecosystem, such as the forested watershed at Hubbard Brook in northeastern America (Borman and Likens, 1967). It is already possible to draw a number of deductions from this work regarding the effects of nutrients and nutrient flow on ecosystem function, as is indicated.

Disturbance of Cycling Processes

The different toxic substances which may be introduced into the biogeochemical cycling processes of ecosystems in several ways may have various very serious consequences. The most commonly encountered of these arises from their *selective concentration* as they pass along the nutrient transfer pathways of the ecosystem. Radionuclides or pesticides which are not toxic at low levels may become concentrated in particular organisms up to a thousand or two thousand times, sufficient to impair or inhibit essential metabolic processes. This matter is discussed further in this text.

Before concluding this review of fundamental aspects of the ecosystem it is necessary to introduce one further idea which is treated more fully in later chapters—that of *succession*.

Succession

The concept of *succession* represents one of the earliest ideas developed in ecology. It explains a process which is both simple and readily observed. In any natural ecosystem it is possible to recognize sequential changes in population representation in communities. Starting from the first *pioneer* populations which colonize any unoccupied habitat, whether this is bare ground, open water, or rock, it proceeds to the populations represented in the final stages beyond which there is no further change, the so-called *climax* communities (Whittaker, 1953). This replacement of one population by another in a continual series from the pioneer to the climax communities is known as *succession* (Fig. 1-15).

Unfortunately succession is by no means as formalized a process as was once considered. For example few ecosystems develop from virgin territory, forming what was once called a *primary succession*. Patches of open ground available for colonization are rarely encountered, being found mainly in landslides, fresh alluvial islands in rivers, and on coastal storm beaches. Most successions begin on sites where existing populations have been degraded, as for example on abandoned agricultural fields or in forest clearings. These are often called *secondary successions*. Moreover, the successional stages themselves intergrade in a continuous process, and it is largely an arbitrary exercise to attempt to separate these into a series of discrete successional steps. Climax communities on closer investigation prove to be rarely stable in time. It is a question of what are sometimes called *proximate* and *ultimate* influences. All too often when apparently final stages of succession are examined critically they are found not to be a true climax. They are held in a climax steady state through the operation of a particular abiotic factor, such as a high water table in a swamp, or by a biotic one, such as grazing on a prairie or man-made fires in a forest.

Nevertheless when the general concept of succession is examined as

Ecosystems

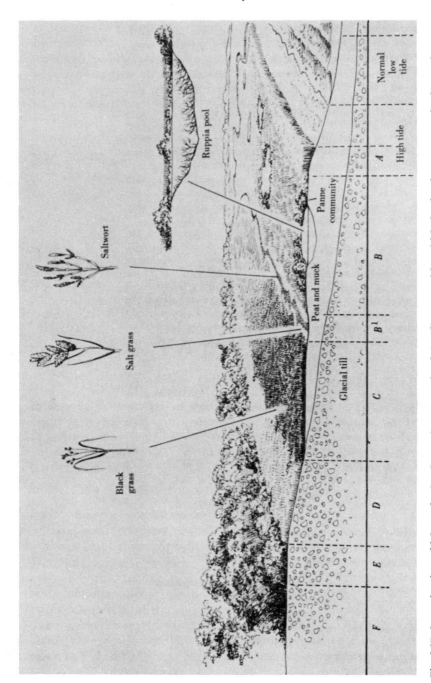

Fig. 1-15. Succession in a tidal marsh, showing various successional (seral) communities which develop in an estuary in the northeastern United States. Each community interacts with and modifies the environment. It is then replaced by another community, until finally the climax communities are established. (From *Ecology and Field Biology* by R. L. Smith, illustrated by Ned Smith: Harper & Row, 1966.)

a whole, certain features appear to be essentially related with successional development. There is an increasing structural complexity of the constituent communities, and an increasing diversity in their populations, a decreasing net productivity and a decreasing dominance. Whereas the early successional stages have an extensive turnover of biomass and an active recycling of nutrients, the climax stages in a given ecosystem are characterized by a balance between internal respiratory processes and the gross productivity of the producer trophic level. The amount of recycling within such a climax community is minimal because there is little biomass turnover. The formation of any waste matter by excretion and by death or loss of portions of organisms starts a reducing cycle which very rapidly decomposes such waste to its basic nutrients. These are then promptly recycled into the unavailable nutrient pools of the ecosystem.

Such climax communities can be utilized for the creation of agricultural ecosystems, but their use can be only temporary, as in the practice known as *shifting cultivation*. Once accumulated biomass has been reduced to essential nutrients which are taken up and removed in agricultural crops, or leached out of the ecosystem, the level of productivity is lowered so excessively as to make the continuation of farming operations uneconomic. Further agricultural use of the land is therefore dependent on a prolonged resting period during which the natural ecosystems build up by succession to climax communities again, or else on the provision of nutrients in the form of artificial fertilizers (Commoner, 1968).

Eutrophication

Bodies of water such as lakes, like terrestrial systems, can be arranged in successional series. Starting from *oligotrophic* waters of low productivity, these graduate to a *eutrophic* condition with high nutrient status and productivity. Each stretch of water tends to progress individually through this succession which is known by a separate term *eutrophication*. As in terrestrial ecosystems, there is a gradual progression from pioneer to climax communities which can truly be described as succession, but the generally ephemeral nature of freshwater bodies, geologically speaking, provides some essential differences.

Eutrophication is treated further in the chapter on community ecology. It is of paramount importance in terms of water pollution, and in matters of water resources and conservation (Beeton, 1965).

Succession in Geological Time

The term *succession* usually refers to contemporary time—that is, the various communities and populations considered are those that exist concurrently and contemporaneously. It is possible, however, to regard succession as a *geological* process, or rather as taking place in geological time. As one dominant form passes to extinction and a new dominant evolves there is a continuous succession of populations, which has been proceeding since life first originated on this planet an estimated three and one-half billion years ago. The pioneer populations which formed the earliest ecosystems would appear to have been autotrophs. Such early ecosystems could not have been self-perpetuating, because no reducer trophic levels would have been present to recycle nutrients to this producer element. Once decomposer heterotrophs had evolved, the ecosystems would become self-perpetuating. It is believed that it took another three billion years before heterotrophs at a consumer level—that is, animals—evolved. There is a considerable body of evidence to support the contention that the earliest ecosystems were marine, and that terrestrial ecosystems were not formed until many millions of years later. Even now the greater variety of plant, animal, and saprobic microbial forms, and the most extensively evolved and diversified ecosystems are encountered in aquatic situations. Curiously, although marine ecosystems had this long start in time over terrestrial ones, they did not evolve creatures anywhere approaching humans in mental equipment. Even cetaceans (whales and dolphins) which come closest, appear to have been terrestrial forms secondarily adapted to an aquatic habitat.

Each geological period is characterized by particular forms of dominant producer and consumer organisms. Thus we speak of the "coal age," or the "dinosaur age," and record the changes that took place as dominant heterotrophs evolved from an amphibian to a reptilian type and finally to a mammalian stage. From being a mixture of such ancient forms as horsetails and cycads, vegetation passed through a predominantly coniferous phase to that of the flowering-plant dominant which characterizes the majority of the world ecosystems today.

This ecological succession in geological time can be regarded as a successive increase in information content in ecosystems. There is no climax in such a geological succession because information never ceases to be added to the system. As information accumulates, the ecosystems are further adapted and therefore further evolve. Each evolutionary progression augments the information store and provides for further diversification within the biotic and abiotic parameters of the ecosystem.

The unique feature of the animal genus *Homo* into which our human population is placed, is that species of this genus are able to adapt to this information accumulation by *cultural* means. In all other organisms adaptation proceeds by the selection of particular variants of the physiological, morphological or behavioral range encountered in given populations. For something approximating to three or four million years little further adaptation has occurred in the genus *Homo* in respect of such physiological, morphological, and behavioral characters. Adaptation has been almost entirely in respect of ability to acquire, process, and disseminate information. The massive technology which human societies have now variously constructed is designed primarily for the acquisition and utilization of new information. Indeed, new information is now being disseminated at such a speed that we are in serious danger of being unable properly to sort and record it in an orderly form. However, this is only one of our present problems concerning ecological relationships within the world's ecosystems. Some of these concerns have been noted in this chapter. Attention will be directed to others as we proceed.

Human Societies and the Environmental Crisis

Our progress during the last two or three million years by *cultural* rather than *genetical* adaptation has now permitted us to invade all the ecosystems of the world (Cole, 1966). Moreover, we are biased toward entering the earliest successional stages of ecosystems where net productivity is high. We therefore tend to destroy the diversity and stability of climax communities, and favor the resulting earlier successional stages, which lack both diversity and stability. As is discussed further in Chapter 3, increasingly involved cultural procedures have permitted us to escape from many of the regulatory feedback mechanisms which once prevented the overexploitation of ecosystem resources.

We have also interfered with ecosystems in many other ways (Fig. 1-16). We disrupt biogeochemical cycling processes by carrying away too much of the nutrients required for ecosystem maintenance. At the same time we dump into the nutrient cycles far more wastes than the reducer organisms can cope with. Or we pour in nondegradable substances to which these decomposers had never previously before been exposed (Cole, 1964; Egler, 1966; Woodwell, 1967). The magnificent achievements of our technology have by their very success pushed us into an apparently inevitable spiral of pollution. Already our eyes smart from smog and our health is imperiled by it. If we try to reduce it however we increase water pollution and intensify its hazards. The endless search for quality of life has already caused us to destroy the natural habitats whose preservation is as vital to our own survival as it is to that of the wildlife populations we have already driven to the point of extinction. Our cities, which have become established as the focal points of our burgeoning civilization (Table 7-1) uncontrollably exploit, deplete, and poison the remaining resources of the planet with an abandon which is beyond all rationality. In our personal relationships within the cities we like to

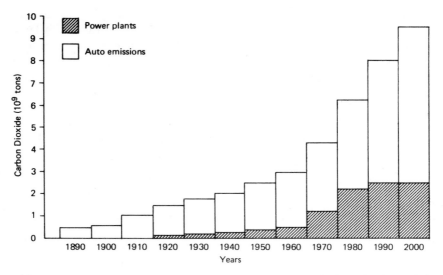

Fig. 1-16. The formation of carbon dioxide in industrial activities, still increasing dramatically, provides an excellent example of human disturbance of natural ecosystems. The recorded and estimated global release of carbon dioxide by industry and automobiles during this century will have risen approximately 20-fold, as shown here. The possible results of this spectacular rise are discussed in the text.

pretend that we are no longer animals, and can safely ignore the ecological relationships which all social animals have evolved, and which are essential for normal social development and peaceful existence.

In subsequent chapters we will examine more closely, and within more restricted fields, the numerous crises that have resulted from our actions in ignoring many implications of the fundamental concepts of the structure and function of ecosystems developed here. The next chapter deals with the effects of abiotic factors. Many of these have extrapolations to more specific environmental consequences.

Bibliography

References

Bakuzis. E. V., "Structural Organization of Forest Ecosystems," *Proc. Minn. Acad. Sci.* 27:97–103, 1959.

Bormann, F. H., and Likens, G. E., "Nutrient Cycling," *Science* 155:424–429, 1967.

Chapman, S. B., "Nutrient Budgets for a Dry Heath Ecosystem in the South of England," *Jour. of Ecology* 55:677–689, 1967.

Cole, LaMont C., "The Ecosphere," *Scientific American* 198 (4):83–92, 1958.

Englemann, M. D., "Energetics, Terrestrial Field Studies, and Animal Productivity," *Advances in Ecological Research* 3:73–115, 1966.

Golley, F. B., "Energy Dynamics of a Food Chain of an Old-Field Community," *Ecological Monographs* 30:187–206, 1960.

Harper, J. L., "The Role of Predation in Vegetational Diversity," in *Diversity and Stability in Ecological Systems,* Brookhaven Symposia in Biology No. 22:48–62, 1969.

Kozlovsky, D. G., "A Critical Evaluation of the Trophic Level Concept. 1, Ecological Efficiencies," *Ecology* 49:48–60, 1968.

Margalef, R., "On Certain Unifying Principles in Ecology," *American Naturalist* 97:357–374, 1963.

Odum, E. P., *Fundamentals of Ecology.* 2nd ed. Philadelphia: Saunders, 1959.

———, "Relationships between Structure and Function in the Ecosystem," *Japanese Jour. of Ecology* 12:108–118, 1962.

———, "The Strategy of Ecosystem Development," *Science* 164:262–270, 1969.

Olson, J. S., "Energy Storage and the Balance of Producers and Decomposers in Ecological Systems," *Ecology* 44:322–331, 1963.

Ovington, J. D., "Organic Production, Turnover and Mineral Cycling in Woodlands," *Biological Reviews* 40:295–336, 1965.

Patten, B. C., "An Introduction to the Cybernetics of the Ecosystem: The Trophic-dynamic Aspect," *Ecology* 40:221–231, 1959.

Phillipson, J., *Ecological Energetics.* New York: St. Martin's, 1966.

Raymont, J. E. G., "The Production of Marine Plankton," *Advances in Ecological Research* 3:117–205, 1966.

Ryther, J. H., "Geographic Variations in Productivity," in M. N. Hill (ed.), *The Sea.* London: Interscience, (1963), Vol. 2, pp. 347–380.

Slobodkin, L. B., "Energy in Animal Ecology," *Advances in Ecological Research* 1:69–101, 1962.

Steeman, N. E., "Productivity, Definition and Measurement," in M. N. Hill (ed.), *The Sea.* London: Interscience, 1963, vol. 2, pp. 129–164.

Turner, F. B., (ed.), "Energy Flow and Ecological Systems," *American Zoologist* 8:10–69, 1968.

Westlake, D. F., "Comparisons of Plant Productivity," *Biological Reviews,* 38:385–425, 1963.

Whittaker, R. H., "A Consideration of Climax Theory. The Climax as a Population and Pattern," *Ecological Monographs* 23:41–78, 1953.

———, "Forest dimensions and production in the Great Smoky Mountains," *Ecology* 47:103–121, 1966.

———, *Communities and Ecosystems.* New York: Macmillan, 1970.

———, and G. M. Woodwell, "Dimensions and Production Relations of Trees and Shrubs in the Brookhaven Forest, New York," *J. of Ecology* 56:1–25, 1968.

Woodwell, G. M., "The Energy Cycle of the Biosphere," *Scientific American* 223 (3):64–74, 1970.

Further Readings in Human Ecology

Beeton, A. M., "Eutrophication of the St. Lawrence Great Lakes," *Limnology and Oceanography* 10:240–254, 1965.

Brown, H., "Human Materials Production as a Process in the Biosphere," *Scientific American,* 223 (3):196–208, 1970.

Cole, LaMont C., "Pesticides: a Hazard to Nature's Equilibrium," *American Journal of Public Health* 54 (1, pt. ii): 24–31, 1964.

———, "Man's Ecosystem," *Bioscience* 16:243–248, 1966.

Committee on Resources and Man, *Resources and Man.* San Francisco: Freeman, 1969.

Commoner, B., "Nature Unbalanced: How Man Interferes with the Nitrogen Cycle," *Scientist and Citizen* 10 (1):28, 9–12 (January–February 1968).

Egler, F. E., "Pesticides in Our Ecosystem," *Ecology* 47:1077–1084, 1966.

Hutchinson, G. E., "The Biosphere," *Scientific American,* 223 (3):44–53, 1970.

Singer, S. F., "Human Energy Production as a Process in the Biosphere," *Scientific American* 223 (3):174–90, 1970.

Woodwell, G. M., "Toxic Substances and Ecological Cycles,"*Scientific American* 216: (3):24–31, 1967.

Review Questions

1. Define an ecosystem; what are its essential features?
2. What external sources of energy are available for the support of ecosystems? What different types of ecosystems utilize these?
3. Describe the process of energy flow within ecosystems.
4. Discuss the ecological phenomena associated with the pyramid of numbers.
5. State what is meant by the pyramid of energy. Explain and discuss the reasons for the introduction of this concept.
6. Describe the pyramid of biomass. In what ways does this concept facilitate an understanding of ecosystem functions?
7. Define productivity. Why are productivity studies vital in ecological investigations?
8. Describe the relationship between primary and secondary productivity. What is meant by the terms "gross productivity" and "net productivity"?
9. What is a trophic level? Describe any differences which occur in energy availability at various trophic levels.
10. Define "food chains" and "foodwebs." What are the particular energy relations within these, and what general consequences are attributed to these relations?
11. In what essential features would you expect a terrestrial ecosystem to differ from a marine one?
12. What are the principal interfaces in an ecosystem where interactions between the components of its biotic element occur? Describe the ways in which these differ.

2

Environments

As has been described in the previous chapter, the abiotic element of ecosystems comprises the totality of physical factors and chemical substances which interact with the biotic element. The abiotic element therefore includes the limiting factors and the limiting nutrients which modify or regulate the dispersal, growth, and reproduction of the component populations of particular ecosystems. The totality of these influences may be described as the *environment*. The study of abiotic factors and substances can be called *environmental ecology*. When it concerns agricultural ecosystems, it is studied by agronomists. If it involves effects on populations of domestic animals, it is investigated by workers in *animal husbandry;* if it relates to cultivated plants, it is examined by *crop physiologists*. Strictly speaking, the abiotic element of an ecosystem is not exactly synonymous with the *environment,* because in addition to the physical and chemical features of the habitat described by this term, it also includes any *biological* interactions such as those considered in Chapter 5. Recently a further group of specialists, the *biometeorologists,* has organized the new discipline of biometeorology to research into the environmental complex.

Limiting Factors

It was in relation to agricultural ecosystems that the concept of *limiting factors* was first developed. Justus Liebig in 1840 observed that crop yield is dependent on the amount of nutrient which is present in the least quantity. This observation subsequently became known as *Liebig's law of the minimum* (Fig. 2-1). At the beginning of the present century this law was extended by F. F. Blackman to include abiotic factors in the more narrow meaning of the word, as well as nutrients. Blackman demonstrated experimentally that the rate at which photosynthesis proceeds is governed by the level of the particular factor which is operating at a limiting intensity (Fig. 2-1).

Subsequent work on limiting factors and substances has shown that their operation is not so precise as was once believed. Factors and substances *interact,* so that high levels, or high concentrations of one substance will to some extent offset low levels or low concentrations of a limiting nature in another. The statistical technique of *multivariate analysis* has been developed to calculate individual effects of limiting factors and substances, both when acting alone and when in combination with other factors and substances. Such factor analysis is a complicated procedure, and is generally applied only to agricultural experimentation and in laboratory simulation studies. Environmental ecologists have preferred wherever possible to *avoid* such a procedure by selecting situations in which all but one factor or substance could be held constant. The systems approach, continuous telemetering devices, and the increasing sophistication of computer techniques all now encourage environmental ecologists to explore more complex and less controlled situations.

Shelford's Law

An important addition was made to the Liebig-Blackman law of the minimum in 1913 by V. E. Shelford. He pointed out that populations could be limited in their growth and reproduction by *too high* a level of operation of a particular factor as well as too low, and *too much* of an essential substance as well as too little. This idea is embodied in Shelford's law of tolerance, which holds that each factor of the abiotic element has a maximum and a minimum level for each constituent population of a given ecosystem, and that between these points lies a range known as the *limits of tolerance* (Fig. 2-2). A considerable mass of data has now been assembled on the tolerance limits of both natural populations and those of domestic animals and cultivated plants. Some of this work is reviewed in a brief survey of the major categories of limiting factors provided in the next section of this chapter.

Reference has already been made to nutrients essential for population growth and development, to the preparation of nutrient budgets to chart the cycling transfer rates and transfer patterns of these nutrients in ecosystems, and to the existence of available and unavailable pools of such substances (Likens et al., 1967). The various chemical materials essential for populations may conveniently be divided as already noted into two classes, *macronutrients* and *micronutrients* or *trace elements.* The former includes molecular oxygen, carbon dioxide, and various compounds of nitrogen, hydrogen, phosphorous, potassium, calcium and sulfur. The

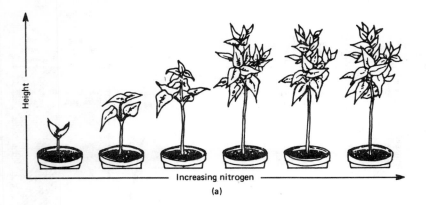

Increasing nitrogen

(a)

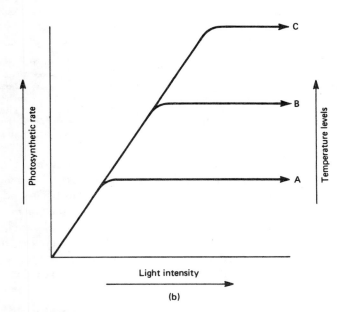

Light intensity

(b)

Fig. 2-1. Limiting quantities and levels. a) Liebig's law of the minimum supposes that the amount of a limiting abiotic *substance*, in this instance nitrate,
can determine production, despite the necessary operation of other factors
which however do not happen to be limiting. b) Blackman's law of limiting
factors postulates that an abiotic *factor*, in this instance light intensity, can
similarly determine the rate of an ecological process. Both laws have subsequently been modified to provide for the synergistic effects of interaction
with other substances and factors.

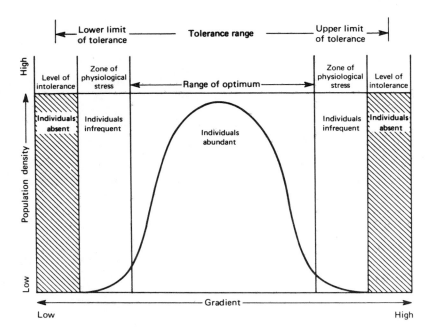

Fig. 2-2. Schematic illustration of Shelford's law of tolerance. The marginal dispersal area of a population is determined partly by its tolerance range for a given abiotic factor. Where physiological stress occurs as a result of exposure to the lower or upper limit of tolerance, any prolonged persistence of such conditions is likely to cause local extinction of members of this population. The main dispersal area of the population therefore tends to coincide with the distribution of microenvironments falling within the physiological tolerance limits.

latter are most commonly boron, manganese, magnesium, zinc, iron, copper, molybdenum, iodine, chlorine and cobalt salts. Micronutrients are utilized in much smaller quantitites than macronutrients, sometimes even in extremely minute amounts, hence the name *trace element*. Macronutrients are needed in considerable quantities and their circulation through ecosystems in the process of biogeochemical cycling has been extensively studied, especially as we have seen in agricultural systems. It is convenient to depict the cycling process for particular essential macronutrients as a series of individual cycles.

The Nitrogen Cycle

Because of the significance of nitrogen as a limiting macronutrient in all agricultural and many natural ecosystems, the nitrogen cycle which is illustrated diagramatically in Fig. 2-3, is the best known and most completely studied of the individual biogeochemical cycles. Atmospheric

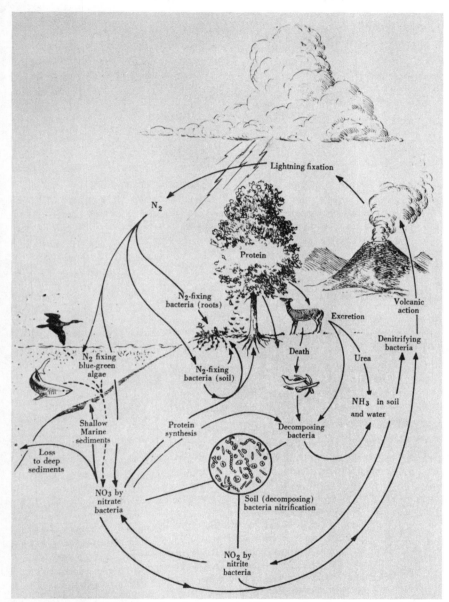

Lightning fixation

N₂

Protein

N₂-fixing
bacteria (roots)

Volcanic
action

Excretion

Denitrifying
bacteria

N₂ fixing
blue-green
algae

Death

N₂-fixing
bacteria (soil)

Urea

Protein
synthesis

Shallow
Marine
sediments

NH₃ in soil
and water

Loss
to deep
sediments

Decomposing
bacteria

NO₃ by
nitrate
bacteria

Soil (decomposing)
bacteria nitrification

NO₂ by
nitrite
bacteria

Fig. 2-3. Diagram illustrating the nitrogen cycle. The size of each nitrogen pool can be estimated in kilograms per square meter, and transfer rates in kilograms per square meter per year (see Bowen 1966). Harvesting of excessive portions of the biomass, as in a forest clear-felling operation, can lead to depletion of the total nitrogen pool, and a lowering of the productivity of the ecosystem. (From *Ecology and Field Biology* by R. L. Smith, Illustrated by Ned Smith: Harper & Row, 1966.)

nitrogen, although it constitutes approximately 78 percent of the earth's atmosphere, is not directly available to populations of the producer-trophic level in any major autotrophic ecosystem. With the exception of the few heterotrophic ecosystems which are able to utilize atmospheric nitrogen directly (Stewart, 1967) it therefore represents an "unavailable" pool. One group of organisms which *can* utilize atmospheric nitrogen are soil bacteria in the genus *Rhizobium*. The rhizobia occur in association with nodules which form on the roots of leguminous plants (e.g., alfalfa, peas, clover, beans, vetches). This same type of association occurs, but less extensively, between other groups of flowering plants like alders, and other microbes. In addition free-living nitrogen fixing bacteria like species of *Azotobacter* are found in many soils. Some blue-green algae are also able to synthesize organic nitrogen compounds from molecular nitrogen.

The synthesis of molecular nitrogen into organic nitrogenous compounds by the fixation of atmospheric nitrogen represents one stage of the nitrogen cycle. The other processes in which nitrogenous compounds are further cycled take place in soil, *mineralization* and *nitrification*. Mineralization is the conversion of the nitrogenous substances in dead organic material into ammonia or ammonium salts, often also called *ammonification*. Many fungi, actinomycetes, and bacteria are capable of this conversion, and the ammonia and ammonium compounds formed represent their way of eliminating excess nitrogenous substances. The ammonia released in this process may be detected, for example, in old horse manure.

Other saprobic soil microbes, more especially nitrifying bacteria belonging to the genera *Nitrosomonas* and *Nitrobacter,* oxidize ammonia and ammonium salts to nitrite, and nitrite into nitrate respectively. In this process of nitrification various saprobes or reducers utilize energy derived from dead organic matter and obtain oxygen by reducing carbon dioxide or bicarbonates. Denitrifying bacteria have the ability to reduce nitrates to release molecular nitrogen and so reverse this portion of the nitrogen cycle.

In aquatic ecosystems some loss of nitrogen compounds occurs not only by denitrification but also by sedimentation. Taking the biosphere as a whole, however, these total nitrogen losses balance the gain by fixation of gaseous nitrogen, nitrification, volcanic release of ammonia, oxidation of atmospheric nitrogen by lightning, and the weathering of rocks. In agricultural ecosystems, one effect of human exploitation is the large-scale removal of nitrogenous compounds in the form of proteins in the seeds or fruits harvested, and in other above-ground portions which are consumed such as straw, hay, and silage. The same occurs when underground storage organs are cropped, such as rhizomes, corms, and tubers as in potatoes, onions, and sweet potato respectively. Animal proteins are also removed in such products as milk and fresh eggs. The consequent losses to the nitrogen cycle are usually made good by the addition of natural manure or artificial fertilizers such as ammonium sulfate.

In temperate regions the inclusion in the agricultural rotation of a leguminous crop such as clover, with its associated rhizobial nitrogen-fixing bacteria, helps to maintain the pool of available nitrogen in the ecosystem. It is estimated that 500 lb of nitrogen per acre per year can be added to a New Zealand pasture by symbiotic nitrogen fixation. Free-living nitrogen-fixing soil bacteria have been determined as adding some-what less, about 100 lb per acre per year.

This problem of maintaining the level of nitrogen circulating in the ecosystem is examined further in several of the volumes in this series. Items discussed there include the extent to which human populations may be estimated already to have risen in some areas beyond the point at which present resources of inorganic nitrogen are sufficient to prevent nitrogen becoming a severe limiting factor (Brown, 1967). This is a central consideration in several of our environmental problems.

The Phosphate Cycle

The other macronutrient most commonly in limiting supply in both terrestrial and aquatic ecosystems is *phosphate*. Whereas the nitrogen cycle belongs to the category sometimes described as a *gaseous cycle,* the phosphorus cycle belongs to the class known. as *sedimentary cycles* (Fig. 2-4). Phosphorus is an important essential macronutrient, because it is a constituent of nucleic acids, phosphorylated compounds and phospho-lipids. It also tends to be recirculated very rapidly in ecosystems following mineralization and nitrification of dead organic matter, as will be described.

There is additionally an "unavailable" pool of phosphates in sedi-mentary rocks. When these rocks are weathered, and the phosphates from them carried down in drainage systems to the oceans, they are again deposited in sediments, some over the continental shelves others in deeper water (Goldberg, 1963). The sediments overlying the continental shelves may subsequently be uplifted by mountain-chain building, and the ero-sion cycle restarted. The phosphate which is carried into deeper water is lost, so that the total pool of potentially utilizable phosphorus in the world appears to be a diminishing quantity. Limited quantities of phos-phates have in the past been obtained from guano deposits, sometimes many feet thick, where over the centuries marine birds have established breeding colonies on small islands. Fish-eating birds also help to recycle

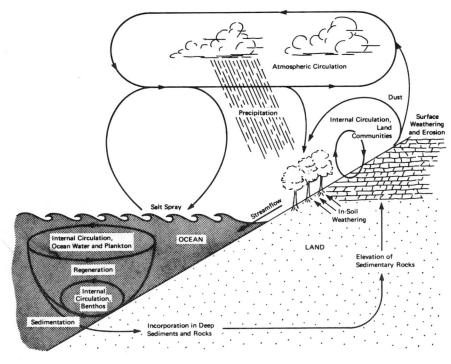

Fig. 2-4. A generalized sedimentary cycle. Various essential elements such as calcium, phosphorus, and potassium are recycled into terrestrial or marine ecosystems by biogeochemical cycling processes essentially similar to that represented schematically here for phosphate. (Based partly on Whittaker, 1970.)

phosphorus in marine ecosystems by returning waste products directly into the seas of their fishing areas.

Mention has already been made of the selective concentration of nutrients and its significance in the contamination of aquatic ecosystems by various poisons and pollutants. Phosphorus under some circumstances is subject to the same selective concentration, as has been demonstrated by Kuenzler (1961). He investigated the cycling of phosphorus in mussel populations established in salt marshes on the southeastern coast of the United States. The food of these shellfish is organic detritus which they filter from sea water. This is rich in phosphorus as well as certain other organic materials, and the effect of this mussel activity on phosphorus concentrations is illustrated in the diagram in Fig. 2-5. The high productivity of coral reefs despite their frequent occurrence in oligotrophic (nutrient poor) waters, is due in part to a similar selective concentration of phosphorus.

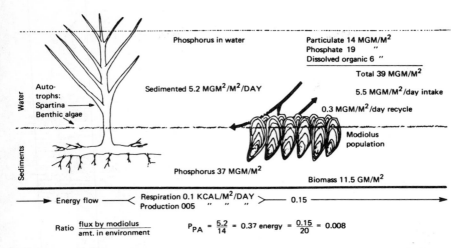

Fig. 2-5. Recycling of phosphorus in a marine ecosystem. The mussels (*Modiolus*) ingest the phosphorus-rich particulate matter of this estuarine habitat. The phosphorus is then recycled in the ecosystem with a turnover time as low as 2.6 days. (After E. J. Kuenzler, *Limnology and Oceanography* 6:400-415, 1961 and E. P. Odum, Japanese J. Ecol. 1962. Reproduced with permission of the publisher.)

Phosphorus in Experimental Ecosystems

The radioisotope of phosphorus, P[32], with a half-life of 14 days, is comparatively harmless and is therefore extensively used to study nutrient budgets and the circulation of materials through simulated ecosystems in laboratory experiments. Such experimental ecosystems are usually aquatic and of the general form illustrated in Fig. 2-6. There is a producer level of some unicellular planktonic algal forms, a herbivore or primary consumer, usually water fleas (*Daphnia*), and a carnivore or secondary consumer represented by a fish species that will feed on the water fleas which graze the algae (Whittaker, 1961).

When a small amount of radiophosphorus is introduced to such an experimental ecosystem, it is taken up rapidly by the planktonic organisms. Within a period of about 12 hours an equilibrium has been reached in the distribution of the isotope as between the water and plankton. Somewhat later, radioactivity can be detected in the water fleas, and then in the fish.

Simultaneously with this movement of radioactive phosphorus from the water through the producer and the two consumer trophic levels, it is also being deposited in the sediment at the bottom of the holding tank in which the experiment is being conducted. By the end of a month the major portion of the remaining radioactivity from the P[32] is usually found to be distributed in this bottom sediment. The reason for this

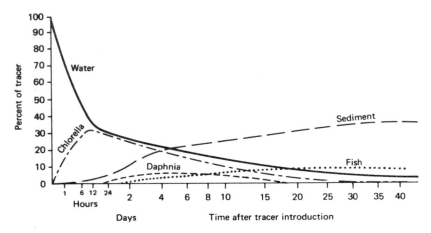

Fig. 2-6. A synthetic ecosystem. Designed for laboratory experimentation, the ecosystem illustrated schematically here is composed of one producer population (*Chlorella,* a unicellular alga) one primary consumer (herbivore) population (*Daphnia,* water flea), and one secondary consumer (carnivore) a fish population. Its decomposer trophic level is composed of many saprobial populations accidentally or inadvertently introduced into the ecosystem, and located especially in the sediment which settles to the bottom of the container. The P-32 tracer is concentrated in first one then another trophic level as it circulates through the ecosystem after being introduced into the water.

accumulation in the sediment is that not all the phytoplankton cells are consumed by the water fleas, which likewise are not all consumed by fish. Dead fleas and fish and the feces of these two consumer levels also sink to the bottom. The reducer organisms which develop in the sediment do not return the radioactive isotope to the water in the form of inorganic phosphates directly and completely, because some of this remains adsorbed on sediment particles; a proportion of the phosphorus remains also in undecomposed organic remains.

The concept discussed in this text of an "unavailable" or "reservoir" pool of a micronutrient and a second and "available" pool has been applied to such an experimental situation. The total amount of the macronutrient, in this instance inorganic phosphate, which is available is known as the *pool content,* and the rate at which it is removed from the system and rendered at least temporarily unavailable in sediments is known as the *transfer rate.* Experimental systems have been quite extensively used for studies of pool content and the transfer rates and patterns of particular nutrients. Some idea of the relative availability of essential nutrients in such pools under natural conditions may be obtained from Table 2-1.

The time which it takes for a chemical substance to pass through a nonavailable to an available pool is sometimes described as the *residence*

TABLE 2-1

Phosphate turnover time in various marine ecosystems, at approximately the same latitude and longitude.

Locality	Month of Year	Turnover Time, Hours
Gulf stream (open water–surface)	July	4
Gulf stream (open water—50-m depth)	July	12
Coastal (open water)	April	4
Coastal (open water)	October–November	46–155
Doboy Sound, Ga. (open water)	February	50
Akamaka River, Ga. (estuary)	May	1
Salt-marsh creek, Ga. (estuary)	July	1
Salt-marsh (low tide)	January and October	40–49
Salt marsh (high tide)	November	169

Adapted from L. Pomeroy, *Science* 161:1731–1732, 1960.

time. The *turnover rate* or *transfer rate* is the rate at which an available nutrient is being transferred to a "nonavailable" pool. The period which elapses before all a given nutrient is so transferred at this rate is the *turnover time.* Mathematical expressions have been developed to define turnover time. *Residence time* is sometimes used as synonymous with turnover time, and likewise is determined by the size of the "available" pool divided by the amount of nutrient utilized in unit time. However, *residence time* can also be applied to the utilization of nutrients from a "nonavailable" pool. Residence times vary from somewhere in the region of 3×10^8 years for atmospheric nitrogen to a matter of a few months in the case of inorganic nitrogen salts in the soil. The same concept of residence time may be applied to the turnover of organic matter as represented by leaf litter which accumulates on the surface of soil. The turnover rate in this instance is inversely correlated with temperature and related to other abiotic factors such as the soil reaction and the source of the material.

In natural aquatic ecosystems it is believed that the turnover time, from the moment mineral phosphate becomes available to when it is stored in an unavailable pool, is very rapid and reckoned in freshwater systems in terms of minutes. In marine ecosystems it takes a little longer, generally several hours (Table 2-1).

Phosphates and Pollution

Phosphorus together with nitrogen represents the major source of pollution of natural waters and is responsible for the phenomenon known as *eutrophication,* which has already been outlined. Not only untreated sewage, but also treated sewage which has not been specially processed for phosphate removal, releases large quantities of phosphates into natural

drainage systems. So does runoff from agricultural holdings where domestic animals are raised, or where artificial fertilizers have been applied
to crops. An almost equally large amount of phosphate emanates from
the biodegradation of modern domestic detergents, which contain a
large quantity of bound phosphate, varying from about 15 to 60 percent.
The immense scale of pollution of natural waters with phosphate from
these several sources and its effects is indicated by the calculation that the
maturation of Lake Erie, now an estimated 10,000 years old, has been
advanced 150,000 years during this century by such pollution. Indeed
few natural waters around the world remain unaffected by any degree of
pollution. Unless particular steps are taken such as have been described
for Lake Washington (Edmondson, 1968), the effects of this pollution
and eutrophication on the aquatic ecosystems involved are disastrous
(Sawyer, 1966; Hasler, 1969).

Other Biological Cycles

There are various other biogeochemical cycles which are of considerable importance in ecology, including the hydrological, oxygen,
calcium, sulfur, silicon, and trace-element cycles. The hydrological cycle
(Fig. 2-7) is critical for human populations, determining not only the
extent of habitable area but also the nature of the economic activity
possible there (Bradley, 1962). The volume in this series on water resources
considers the pools of "available" and "unavailable" water to which
human populations have access, and the limits which these place on
population densities and activities (Table 2-2). These activities have indirect environmental consequences of considerable magnitude arising

TABLE 2-2

Distribution of water over the globe illustrates the very rapid turnover rate for
atmospheric moisture. The total melting of all ice caps and glaciers would increase fresh and marine waters by about 1 percent, and cause a rise of approximately 100 m in sea level. The annual global precipitation represents only about
one-fortieth of this amount. The pool of atmospheric water vapor is even smaller.
It is estimated as sufficient to produce a rainfall of about 25 mm over the whole
globe—enough for just one reasonable rainstorm. Without the hydrological cycle
the whole land surface of this planet would therefore become a desert within a
matter of weeks.

Portion of Globe	Water Content, liters
Rocks and ground water	$250,000 \times 10^{17}$
Oceans and other free water surfaces	$14,000 \times 10^{17}$
Polar ice caps and other glaciers	170×10^{17}
Atmospheric water vapor	0.1×10^{17}

Data from various sources.

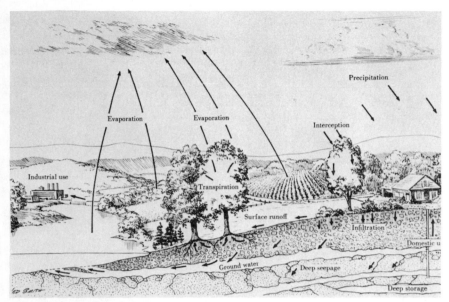

Fig. 2-7. The hydrological cycle. The circulation of water within an ecosystem is controlled by certain physical laws, particularly those relating to surface evaporation and water retention in small spaces. Solar radiation supplies the energy for evapotranspiration. Surface-tension effects combine with gravity to regulate the movement of water in soil. Disturbance of this hydrological cycle, for example by increasing the rate of surface runoff, or pumping from deep storage supplies, may permanently change the nature of the ecosystem and degrade its productivity. (From *Ecology and Field Biology* by R. L. Smith, illustrated by Ned Smith: Harper & Row, 1966.)

from the effects which air pollution has on the local and global patterns of the hydrological cycle (Aynsley, 1969).

The availability of many of the macro- and micronutrients circulated in the sedimentary type of cycle is dependent on continued periods of orogeny—that is, on a continuation of the geological process of uplift and mountain building. The regions where this mountain-chain uplift usually occurs have previously been marine or freshwater sedimentary basins. The sedimentary rocks which formed there contain the various nutrients which have been leached from the soils overlying weathered rocks from a previous orogenous period.

Sedimentary cycles operating on a geological time scale thus ensure that all macro- and micronutrients essential for population growth, aside from those which have a gaseous phase, are recirculated through the ecosphere of this planet (Table 2-3). Such geological processes can be little disturbed by man. This is unfortunately not true of the hydrological and gaseous cycles like those for oxygen and carbon dioxide: the

TABLE 2-3

Salt content of water from various sources—in grams per liter (parts per thousand), of the most common ions. Seawater contains very approximately one hundred times the content of all these salts except carbonates as compared with "hard" potable water. The salt content of rainwater is so low it is more commonly expressed in parts per million.

	Seawater	Domestic Water (Soft to Hard)	Rainwater
Positive Ions			
Calcium	0.4	0.010 –0.065	0.0001
Magnesium	1.3	0.0005–0.014	0.0001
Sodium	10.7	0.016 –0.021	0.0003
Potassium	0.4	0.002 –0.016	0.0001
Negative Ions			
Carbonate	0.07	0.012 –0.120	Trace
Chloride	19.3	0.019 –0.041	0.0005
Sulphate	2.7	0.007 –0.025	0.002

consequences of human disturbance to these cycles have already been stated.

Environmental Factor Interactions

Apart from the limitations placed on the development of ecosystem populations arising from their essential nutrient requirements, the abiotic element of ecosystems as has been noted also includes the many environmental factors that limit or regulate these populations. One of these factors, the rate of energy exchange between the various trophic levels and the environment, is of special concern to the environmental ecologist. This energy-environment relationship has already been considered in regard to the factors which influence primary productivity. Among the other factor interactions between populations and ecosystems, one major influence is that of the composite factor described by the term *climate*.

Climate

The earlier phases of synecology were preoccupied with studies on the interaction between climate and community types (Daubenmire, 1956). One result of these studies was the establishment of a series of broad correlations between the occurrence of particular *biomes* and such climatic factors as the amount and distribution of the mean annual rainfall (Fig. 2-8). Another series of such correlations was the basis for

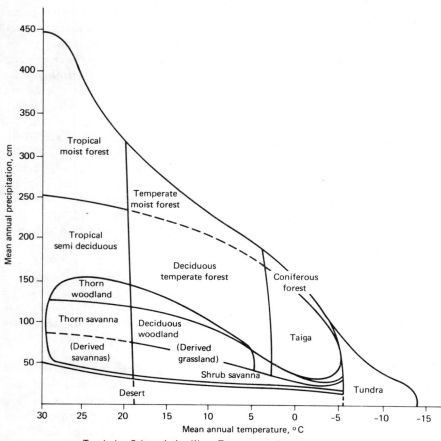

Fig. 2-8. General climate effects. The biome or formation types which are universally employed as convenient major groupings of ecosystems can be characterized by particular moisture and temperature ranges. The above composite scheme, prepared originally by Whittaker (1970) illustrates this broad relationship, and should be contrasted with the microenvironmental effects indicated in Figs. 2-10, 2-12, and 2-13. Disturbance of the hydrological cycle resulting from air pollution is believed to be influencing both the macro- and now the microclimate within many ecosystems.

the classification of the world into climatic zones (Thornthwaite, 1958). Of the several factors which together constitute what is called climate, the one with the most direct ecological effects is *temperature*.

Temperature

The tolerance limits of many species in regard to temperature have now been investigated. The lower limit of tolerance for most species

appears to be less critical than the upper limits. In many organisms there
are devices that appear to modify the effects of the range of temperatures
encountered in the environment. These limit to a much narrower range
the temperatures to which the tissues of the organisms concerned are
actually exposed. In the so-called cold-blooded or *poikilothermous*
(exothermic) animals and in plants, the range of temperatures is nevertheless considerable. In warm-blooded or *homothermous* (endothermic)
animals represented more especially by mammals and birds, the range is
appreciably reduced. In man internal temperatures over 44°C are likely
to prove fatal in a matter of minutes, while unconsciousness and finally
death is reached if the internal temperatures fall significantly below
21°C for prolonged periods. The extensive distribution of human populations about the globe under widely varying prevailing temperatures is
possible because of cultural adaptations, such as clothing and fire. These
permit human populations to modify external temperatures and avoid
their overriding the mechanisms which prevent their body temperatures
from fluctuating beyond these tolerance limits.

There are both horizontal and vertical fluctuations in prevailing
temperature over the earth's surface corresponding to variations in
latitude and altitude respectively. Natural historians still use such
variations for the classification of biome types, and in North America
these are based on the classical zones first established by Merriam in
1890 and illustrated in Fig. 2-9.

One of the early modifications to such correlations between temperature and biological distributions arose from the recognition that
commonly the limiting environmental factor was not so much prevailing
mean temperature as the incidence of short but catastrophic periods of
temperature fluctuations beyond the temperature-tolerance limits of
particular species. Dramatic examples of this are encountered in agriculture and horticulture. In the orange growing districts of America, the
Pacific southwest and Florida, the occasional nights when air temperatures fall below freezing would be fatal to the maturing crop. Various
devices are therefore installed in orange orchards to prevent frost damage
during these occasional catastrophic periods.

Microclimate

Such broad effects of temperature as these are now considered of
lesser importance in ecology than specific local effects at particular surfaces or in more restricted areas (Fig. 2-10). Many modern studies of
microclimates, or strictly speaking of microenvironments, emphasize these
more local temperature effects. Work such as that of Gates (1968) has
stressed the necessity of considering the so-called "boundary" effects of

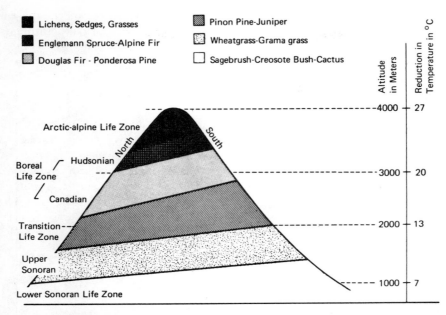

Fig. 2-9. Altitudinal life zones in North America. Just as it is convenient to recognize *biome* or *formation* categories despite the undoubted existence of community gradients, so it is customary to refer to altitudinal life zones, although community diversity likewise tends to vary continuously with altitudinal change. Such altitudinal life zones were originally described and named by Merriam in 1890 and have been subsequently modified according to aspect and locality, in this instance the southern Rocky Mountains.

temperature rather than the broad effects of the macroenvironment, which provide too crude indications for many ecological purposes (Figs. 2-11 and 2-12).

Teleoclimate

Gates has introduced the term *teleoclimate* to include the microenvironmental factors which operate in "boundary" effects. Two definitive papers which he has published describe teleoclimates in relation to animals and plants (Gates, 1968, 1969).

One of the major factors interacting with temperature in the teleoclimate is air disturbance, which also reacts in a synergistic way with other factors such as rate of water loss, which again influences temperature. Population ecologists exploring effects of the teleoclimate are only just beginning to pay sufficient regard to the physical properties of these relationships. The results of such studies on human populations will have an important bearing on the nature of the protective clothing we wear, and the structures in which we live.

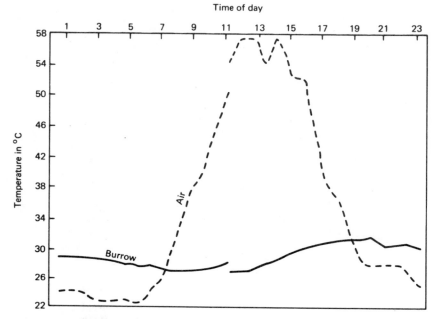

Fig. 2-10. The contrast between macroenvironment and microenvironment. Burrowing animals are able to avoid exposure to environmental factors such as high or low temperatures which exceed their limits of tolerance by utilizing or constructing microenvironments with more restricted ranges of such factors as illustrated in this example supplying temperature data from a gopher burrow. (After T. E. Kennerly, *Texas Journal of Science* 16:4, 1964, reproduced with permission.)

Precipitation

Another very general factor of the macroenvironment is the amount of moisture made available to the biotic element of ecosystems at the various stages of the hydrological cycle, and in particular what is deposited in terrestrial ecosystems as *precipitation*. The commonest form of precipitation is rainfall, but over many temperate regions of the earth in winter, and in the polar regions, snow constitutes the main portion of the precipitation, with many important ecological consequences (Fig. 2-13). The world distribution of biomes as well as more regional distributions show a close correlation with the mean annual amount of precipitation, as is apparent from Fig. 2-8.

There are also important local effects of precipitation, of which *rain shadow* is the most striking. The American deserts of the Great Basin, the Mohave and Sonora are the result of the interruption of rain-bearing winds by the mountain chains (Fig. 2-14). Before the orogenies

Fig. 2-11. The contrasting effects of microclimate—at timberline at just over 3000 meters (10,000 ft.) altitude may be observed on a number of mountain chains such as the Rockies, the Cascades and the Sierra Nevada. Trees like the white fir illustrated here, in winter lie under a blanket of snow about one meter deep. The shoots of the branches under the snow are protected from the damage to their growing points which would otherwise be caused by icy winter winds; they are able to resume normal growth during spring and summer. The shoots projecting out of the snow are extensively damaged during winter storms, especially on the windward side, and their subsequent growth shows the effect of this damage. This phenomenon is commonly known as the *Krummholz effect*.

at the close of the Tertiary Period which raised the Cascade and Sierra Nevada mountains, there were no such Great Basin deserts. Simultaneously with the rain-shadow effect, the exposed sides of the mountain chains which cause them have an exceptionally heavy precipitation with equally dramatic ecological consequences. The "rain forest" of the Olympic peninsula in the State of Washington, lying on the windward side of the Olympic mountains, provides a good example of this.

In the geological history of the earth the level of precipitation has varied very considerably. There have been periods both drier and wetter than at present, but the first have predominated. We appear currently to be enjoying a somewhat cooler and slightly wetter climate than has prevailed over the major part of geological time, and the ecosystems which we have invaded and exploited will be considerably less productive under the warmer and drier climates we may anticipate within the next 2,000 to 10,000 years.

Fig. 2-12. An anomalous effect of microclimate—where the air is permanently saturated with water, as may be observed at the majestic Victoria Falls on the Zambesi River between Zambia and Rhodesia. This region has a six or seven month dry season, but because of the spray which rises continuously from the largest waterfall in the world, an evergreen forest is formed in the spray zone immediately adjoining the falls. Species of plants occur in the "Rain Forest" which are characteristic of the wet tropics of the equatorial zone which lies some 1,500 miles to the north.

As in the case of temperature, ecologists have now turned from considerations of the broad effects of precipitation to more specific and more local ecological effects such as those illustrated in Fig. 2-15. Agronomists have directed their interests in precipitation toward what is generally known as *evapotranspiration.*

Evapotranspiration

Evapotranspiration is the total moisture which evaporates from any specific area of soil and vegetation in a particular ecosystem. It bears a physical relation to the radiant energy which these surfaces receive. Because of this theoretical consideration and perhaps somewhat unexpectedly, a given area of forest loses as much moisture by evapotranspiration as does for example the same area of grassland under the same conditions (Penman, 1963). This equivalence is usually modified in practice by the circumstance that the forest may be evergreen, and therefore loses moisture throughout the year, whereas part or all of the grass-

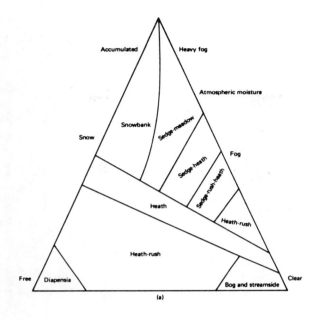

(a)

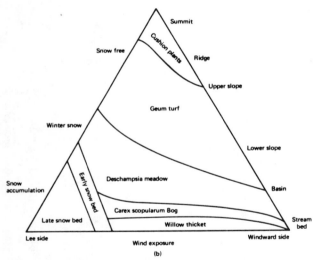

(b)

Fig. 2-13. Microclimatic effects of precipitation. Within the generalization that rainfall tends first to increase, finally to decrease with increasing altitude, there is a pattern of microenvironmental interactions as illustrated schematically here for the Presidential Range of New Hampshire, which rises to a little over 6,000 feet. The influence of precipitation as an environmental factor is related to the form in which it is distributed—(a) mist, rain, or snow, whether it accumulates as a snow-bank or bog—and to certain other factors having local effects like slope (b), as well as to the total amount. (From Bliss, 1962, and Johnson and Billings, 1963, reproduced with permission of the publishers.)

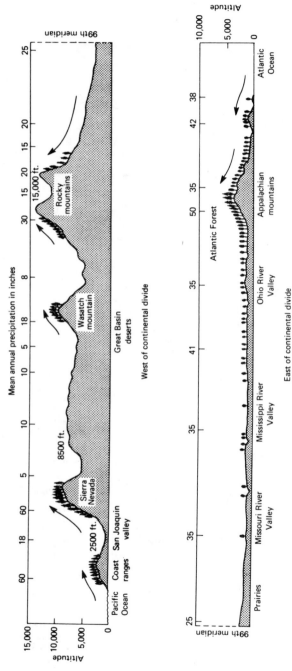

Fig. 2-14. The interaction between topographic and climatic factors. A summarized schematic representation of biome distribution in the United States along the 39th parallel first illustrated by R. Zon in *Climate and Man,* in *Climate and Man, U.S. Department of Agriculture Yearbook 1941.* West of the continental divide, rain-shadow effects produce subdesert conditions in the San Joaquin valley and the Great Basin. East of the divide the lower Appalachian Range has a similar, but lesser effect over the plains of Oklahoma, Missouri, and Kansas.

(a)

(b)

Fig. 2-15. The effect of aspect on communities. a) A south facing chaparral slope on the Irvine Ranch, Southern California. As indicated in Fig. 2-16, the south-facing slopes are covered with an evergreen shrubby vegetation in which dominant rodents include *Peromyscus maniculatus* and *Neotoma lepida*. b) A north-facing chaparral slope, carrying evergreen woodland, as is shown in Fig. 2-16. This contrasts the plant dominants of such north slopes with those facing south as in a). Animal dominants include *Peromyscus californicus* and *Neotoma fuscipes*.

land may die during the inclement winter season, and water is not therefore continuously lost throughout the year.

The amount of evapotranspiration in a given ecosystem may be determined simply by the difference between the precipitation falling on its catchment area, and the amount which runs off as estimated by stream flow over weirs. Under experimental conditions *lysimeters* are employed, containers which prevent incidental water loss so that evapotranspiration may be calculated directly from the amount of water added periodically. Alternatively, the amount of water necessary to maintain the soil water content at a specific level may be recorded, thus providing a figure of the amount of evapotranspiration.

Most studies on evapotranspiration have been carried out on agricultural crops. Only a few attempts have been made to assess this for natural ecosystems. The results of such experiments on an agricultural crop are illustrated in Table 2-4. It is apparent from such experiments as this

TABLE 2-4

The effect of evapotranspiration on agricultural productivity, in spring wheat grown without irrigation. The amount of evapotranspiration is dependent upon the available pool of water stored in the soil from the current and previous season's rainfall. The system of alternate year's fallowing practiced in the more arid regions of the United States reduces evapotranspiration in the fallow year and provides a larger pool of water for crop growth.

Soil Water Pool at Planting Time	Percentage Figure of Maximum Crop Yield
Soil Wet to 1 foot or less	
Soil cropped in previous season	33
Soil *not* cropped in previous season	35
Soil Wet to 2 feet or less	
Soil cropped in previous season	59
Soil *not* cropped in previous season	63
Soil Wet to 3 feet or more	
Soil cropped in previous season	79
Soil *not* cropped in previous season	100

Data from various sources.

that, as in most areas of the world, a crop can be grown under conditions where it is subjected to water strain for approximately one half of its growing life. In other words, water may be a limiting factor in the growth of these agricultural producer populations for approximately half their growing time. The earliest farmers selected empirically for agricultural varieties which tolerated such conditions of water strain. Modern agricultural experimental techniques can likewise select varieties which have effective means of restricting water losses during adverse periods without suffering permanent damage or loss of yield.

Several decades ago plant ecologists were very much concerned with biological methods of avoiding water strain and water loss; terms such as *xerophyte* and *xeromorph* were employed to indicate respectively plants that grew under drought conditions and those that possessed characteristics assisting the restriction of water loss from their surfaces. Animal ecologists are still very much interested in the water relations of terrestrial animals (Edney, 1967). An understanding of the fundamentals of the ecological relationships of plant and animal adaptations to the factor of water loss is clearly required in order to explain the dispersal of particular dominant forms such as shrubs with small rigid leaves in chaparral, evergreen trees with simple entire leathery leaves in tropical rain forests, and kangaroo rats in the American deserts. For the present, however, emphasis has shifted to other aspects of ecosystems and more particularly as already noted to productivity studies (Fig. 2-16).

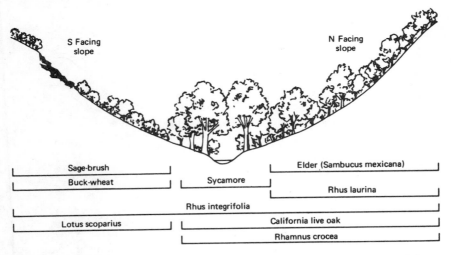

Fig. 2-16. **Topographic effects on microenvironments.** Aspect effects are reported from all the biomes of North America, but the most marked differences occur in the coastal chaparral of Southern California illustrated semidiagramatically here. The microenvironment of the north-facing slopes varies so greatly from that of the south-facing slopes that a different life form (evergreen tree) replaces the typical chaparral life form (evergreen shrub), and the animal communities show parallel differences in population representation. The nature and mode of operation of the various environmental factors involved have not yet been adequately explained.

Climographs

Evapotranspiration is thus a complex ecological factor in which it is possible to recognize the separate influences of several component factors such for example as temperature and moisture. In autecological studies it is usual to express the effects of such interacting factors in the form of what is known as a climograph (Fig. 2-17).

Preparation of such figures is of interest when considering the introduction of new plant and animal species. The tolerance limits for the

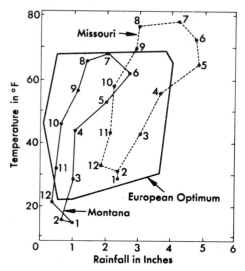

Fig. 2-17. Climograph showing optimum mean European temperature and moisture levels for the breeding of the Hungarian partridge. These are contrasted with similar climographs for Montana (suitable) and Missouri (unsuitable). Introduction of the partridge succeeded in Montana, failed in Missouri. (After A. C. Twooney, *Ecology* 17:122-132, 1936.)

species being introduced can be assessed against the parameters expressed in the climograph that has been prepared for the area into which the species is to be introduced. Climographs are also of significance (although they have not been extensively used in this respect) in warning against the hazards to be encountered when potential weed species of animals or plants might accidentally be introduced to a given area.

Light

The last environmental factor to be considered here is light, which influences ecosystems through variation in its intensity, its wavelength and its duration (Butler and Downs, 1960).

As has already been seen, solar radiation is the source of energy input into the vast majority of autotrophic terrestrial and aquatic ecosystems on this globe; any fluctuations in the value of the incident radiation therefore have a direct effect on these ecosystems through consequent fluctuation in the energy budgets. The proportion of the solar flux which is actually received by terrestrial consumer populations is shown in Table 2-5. The extent to which this fluctuates about the globe is given

TABLE 2-5

The amount of solar radiation utilized by producers in photosynthesis is remarkably low, representing a mere 1 percent of the total solar radiation intercepted by the planet Earth. Sufficient visible light is reradiated from clouds and from the land and water surface to provide a television picture of our planet through the space capsule cameras of voyaging astronauts. Within the solar system this light is believed to provide a brightness roughly comparable with that of Venus. These figures illustrate the obvious possibilities for the introduction of a higher percentage of radiant energy into various ecosystems, with correlated increases in ecosystem productivity.

Amount of Radiation Intercepted	Percent Figure of Total Solar Radiation
By particulate matter, clouds, water vapor and gases such as ozone and carbon dioxide	75
Reradiated or redeflected into space	42
Absorbed	15
Reradiated to earth's surface	18
Amount of Radiation Received at Surface	
By direct radiation	25
By reradiation	18
Total	43
Utilization of Radiation Received at Earth's Surface	
In evaporation	22
In convection	10
In reflection	10
In photosynthesis	1

TABLE 2-6

Variation in solar flux over the earth's surface. Amounts of solar radiation on various plant surfaces received at midday expressed in calories per square centimeter per minute. Data extracted from Lemon et al. 1970. The amount of radiation received is maximized at lower latitudes and the height of summer.

Costa Rica	1.3	(jungle, Nov.)
Japan	1.3	(corn, Aug.)
Australia	1.2	(pine, Nov.)
Germany	1.1	(spruce, Aug.)
U. S. A.	1.1	(corn, Sept.)
Australia	0.9	(wheat, Oct.)
England	0.7	(sugar beet, Aug.)

in Table 2-6. The duration and amount of light may have effects such as that illustrated in Fig. 2-18.

Equally important effects of light on ecosystems are the regulatory ones in which the length of the light period during the day, the photo-

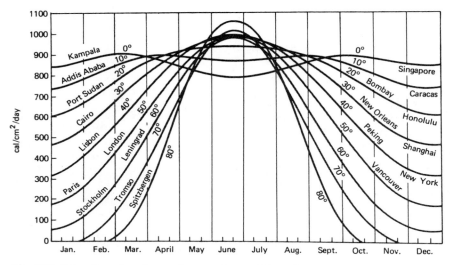

Fig. 2-18. Solar flux at various points on the earth's surface. The daily total of solar radiation which is received at different times of year at various latitudes. (After D. M. Gates, 1962. Reproduced with the permission of the publisher.)

period, triggers phsiological mechanisms which initiate particular phases of the life history of the organisms concerned, in particular the reproductive phase (Beck, 1960). Of the many environmental factors to which animals and plants could relate to delimit the various phases of their life cycle, photoperiod is the most reliable (Fig. 2-19). The relative lengths of day and night are astronomically determined, and are not subject to unpredictable fluctuations. For a population that will be subjected to selection pressures favoring constancy in the timing of the various life-history phases, adaptations associating these phases with particular photoperiods would therefore seem to have selective value. So spawning in fish, mating in mammals, pairing in birds, the onset of flowering and fruiting in plants, and sometimes seed germination, are all responses to photoperiod. The timing of long distance migration in birds and in sea mammals may likewise be influenced by photoperiod responses.

When such photoperiod responses have been investigated, commercial applications have often developed. Whereas we once associated certain cultivated flowers with particular seasons, artificial adjustments

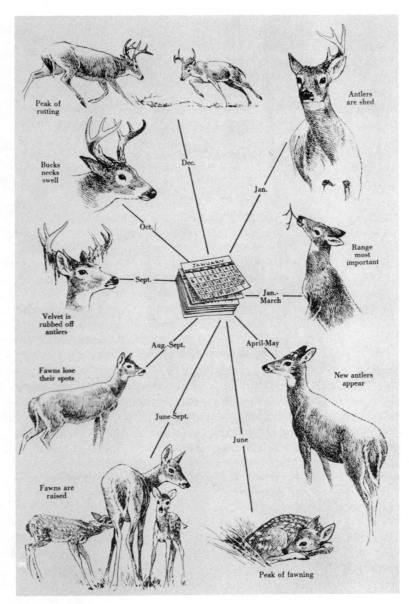

Fig. 2-19. The photoperiod and seasonal activity. The seasonal phases of activities of many animals, as also of plants (*phenology*), are commonly triggered by particular photoperiods. In the case of the white-tailed deer illustrated here, the annual breeding cycle is initiated by the decreasing fall day length. For the purpose of realizing more continuous food supplies over the whole year, the photoperiod to which domestic animals and cultivated plants are exposed is commonly artificially adjusted. (From *Ecology and Field Biology* by R. L. Smith, illustrated by Ned Smith: Harper & Row, 1966.)

of the light period in plants under cultivation, or of the chemical substances which control flowering, has made it possible for roses to bloom in spring and daffodils in summer. Likewise the glut of eggs which developed as hens were exposed to the lengthening hours of spring daylight has now been partially avoided by suitable artificial adjustment of the photoperiod in hen houses. Most of photoperiod response is not in fact a response to the length of *day,* but rather to the length of the *dark* period. It is therefore relatively simple to adjust a short day to a long day by introducing a period of light into the long night to break it into two shorter periods.

Complex Factors

There remain in addition to these factors, numerous others for which space does not permit a review, such as altitude, wind speed, and barometric pressure. There are also a number of environmental factors that do not conveniently fall entirely within the abiotic element of ecosystems, but include features both of this and of the biotic element.

Of these the most complex and the most important is *soil* (Jackson and Raw, 1966). To dismiss the soil factor in a few short paragraphs could be very misleading were it not emphasized that heterotrophic ecosystems in the soil constitute a complex so large and important as to parallel the totality of autotrophic terrestrial ecosystems. Soil sciences such as pedology, the study of soil formation, and soil chemistry represent disciplines in their own right and extensive research in these areas has been undertaken. Nevertheless, much still remains unknown or uncertain.

Soil

While soil may be regarded as an ecological factor and one of the components of the abiotic element of an ecosystem, it is equally possible to consider it as a whole complex or conglomerate of heterotrophic microecosystems, containing many different interacting foodwebs and food chains of reducer organisms (Fig. 2-20). It can also be subdivided into many microenvironmental factors such as soil water content, soil reaction, oxygen content, carbon dioxide pressure, temperature, and nutrient and organic contents (Table 2-7). These form a pedological complex as vital to the function of terrestrial ecosystems as the biometeorological ones (Fig. 2-21). Were it not for the biological activity of the multitudinous populations of saprobes in the various soil heterotrophic microecosystems, there would be no biogeochemical cycling and essential nutrients would remain locked in "unavailable" pools. Some of these aspects are again considered in the final chapter of this text.

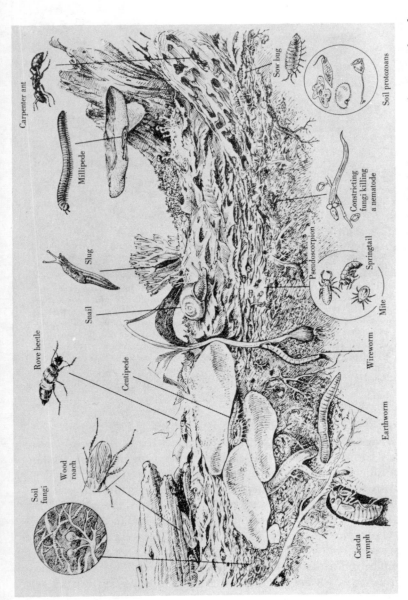

Fig. 2-20. Soil ecosystems. The saprobial communities of the soil provide the food chain bases supporting a complex of microecosystems within the main basically heterotrophic or heterotrophic based soil ecosystem. Although the energy input into these soil ecosystems is in the form of chemical energy, they contain trophic levels of herbivore or carnivore populations, as well as a multitude of saprobial populations at the decomposer level, and parasite populations behaving as herbivores or carnivores. Populations in these ecosystems tend to show the greatest density and diversity in the region immediately surrounding living plant roots, commonly described as the *rhizosphere.* (From *Ecology and Field Biology* by R. L. Smith, illustrated by Ned Smith: Harper & Row, 1966.)

Fig. 2-21. The effect of microenvironment in riparian situations may be observed in
many parts of the world. It is most obvious in regions of seasonal drought where the
presence of an available water table in the valley bottom ensures that evergreen woody
species, and especially trees, can survive along the river bank even though soil conditions elsewhere would be too arid during the dry season for this. The photograph here
shows an evergreen riparian forest along the banks of the Limpopo river which
separates Rhodesia from South Africa. Such microecosystems as this riparian forest
represents may have great ecological significance. For example, disease transmitting
insects such as tsetse fly could potentially survive the inclement dry season in such a
favorable microenvironment.

Pedologists classify soils on the basis of the character of the soil profile. This is divided into soil horizons. In the first, the A horizon, dead
organic matter from natural ecosystems is undergoing *humification*—
that is, conversion to humus, a relatively inert organic stage in the decomposition process. This is underlain by the B horizon in which
mineralization and *nitrification* is taking place. This is the conversion
of the humus and other organic material into its inorganic components,
representing the recycling process of the terrestrial ecosystems in which
the soil is located.

The classification of soil types, and the development of the various
cultural treatments which are applied either to modify certain factors of
the soil microenvironment (such as its water content or its nutrient
content), or to accelerate some function of its heterotrophic microecosystems (such as by drainage), all have a great significance for human

TABLE 2-7

Correlation between topography and soil conditions. While the soil catena concept has most frequently been applied to tropical soils, temperate ones which are particularly susceptible to leaching show well-marked (catena) soil patterns as illustrated by the figures below taken from a classic work on the English chalk "downs." Series A was sampled at depths of 0 to 1 inch, series B at 2 to 4 inches.

| | Percent Carbonates | | Soil Reaction | |
	A	B	A	B
Summit of down	0.00–0.02	0.01–0.04	5.1–5.4	5.3–6.0
Midway down slope	0.68–1.0	1.1 –3.2	7.3	7.3–7.4
Base of slope	2.1 –3.0	2.5 –6.5	7.4–7.6	7.6
Valley floor	12.5	2.0	7.5	7.5

ecology. So do various degradation procedures which result in *soil erosion* and lead to drastic reduction of the productivity of the ecosystems involved. Space does not permit further consideration of such factors here, but some aspects are taken further in various volumes in this series.

Fire

Another major ecological factor of considerable consequence which is not clearly attributable entirely to the biotic or abiotic element of an ecosystem is *fire*. Wild-fires or man-made fires are one of the most direct and drastic of the environmental influences which affect both natural and man-made ecosystems. They are known to have been ocurring for a very considerable time. Fossil remains of vegetation from as far back as the Permian period some 240 million years ago show evidence that it was swept by fires. Outbreaks in contemporary coniferous forests and other types of vegetation have been demonstrated to result from lightning strikes. However, since the discovery of fire and its use by human populations—which is conjectured to have originated a third of a million years ago, accidental or deliberately set fires have vastly predominated over naturally occurring ones.

It seems possible that large stretches of the land surface of the earth, such as the *boreal forest* reaching in a belt across Canada, Northwest Europe and the north temperate portions of Asia, are covered with ecosystems still dominated by coniferous trees rather than some more modern group of tree, because conifers are better able to recover from the devastating effects of forest fires. Whole areas such as the tropical African savanna have evolved in the presence of fire as an environmental factor. So has the chaparral of the Pacific Southwest (Fig. 2-22) and other local biomes in North America. The removal of the fire factor may

Fig. 2-22. The effect of fire may be general, as in this photograph of a burnt-over area of chaparral in Southern California. The chaparral ecosystem has been evolving for thousands of years in relation to lightning set wild-fires as a factor in the ecosystem. All biota therefore show some adaptation to this fire factor.

change the nature of the climax of an area. It is quite likely to result in the dominance of plant and animal forms very different from those occurring in the original so-called fire-climax community.

Fire also played a very vital role in human evolution. Its application to domestic cooking tenderized food, and reduced mortality accordingly in those sections of the population with less efficient dentition such as the young and the old. In providing domestic heat it probably made possible the extension of the distribution of human populations into seasonally cold regions not otherwise tolerable. It made possible a quantum advance in hunting techniques using fire to drive animals in required directions, and produced the first wave of animal extinctions by "overkill". It permitted the development of the so-called shifting cultivation system, the standard technique by which agricultural communities developed tropical agriculture. Although an extensive literature now exists as to the various ways in which plants and vegetation are affected by and overcome the effects of fires (Fig. 2-23), less work has been done on the way in which animal communities and human communities react. Further research urgently needs to be done (Cooper, 1961).

The various chemical substances and physical factors that comprise the abiotic element of ecosystems are therefore of primary importance in

Fig. 2-23. Selective effects of fire where present suggest that its incidence as an ecological factor is a comparatively new one in the ecosystem. In the deciduous forest ecosystem of Southwestern Zambia, which includes Teak (*Baikiaea*) as one of the tree dominants, the incidence of man-made fire has apparently increased greatly during this present century. A number of plants, including the Teak, are not resistant to such fires. They are being progressively and selectively removed from the ecosystem as shown here.

determining the nature and functions of the biotic element. Profound disturbances have now resulted from deliberate or inadvertent human interference with available resources of the one or the natural operation of the other. These have had very serious consequences for human societies in terms of air and water pollution, mineral resource depletion, food production and conservation as is described in the further reading references at the end of this chapter and in the various volumes in this series.

Bibliography

References

Bazilevic, N. I., and L. E. Rodin, "The Biological Cycle of Nitrogen and Ash Elements in Plant Communities of the Tropical and Subtropical Zones," *Forestry Abstracts* 27:357–368, 1966.

Beck, S. D., "Insects and the Length of Day," *Scientific American* 204 (2):108–118, 1960.

Billings, W. D., and H. A. Mooney, The Ecology of Arctic and Alpine Plants, *Biol. Rev.* 43:481–529, 1968.

Bolin, B., "The Carbon Cycle," *Scientific American* 223 (3):124–32, 1970.

Bowen, H. J. M., *Trace elements in biochemistry.* New York: Academic Press, 1966.

Butler, W. L., and R. J. Downs, "Light and Plant Development," *Scientific American* 205(6):56–63, 1960.

Cloud, P., and A. Gibor, "The Oxygen Cycle," *Scientific American* 223 (3):110–23, 1970.

Cooper, C. F., "The Ecology of Fire," *Scientific American* 204(4); 150–160, 1961.

Daubenmire, R. F., "Climate as a Determinant of Vegetation Distribution in Eastern Washington and Northern Idaho," *Ecological Monographs* 26:131–154, 1956.

Delwiche, C. C., "The Nitrogen Cycle," *Scientific American* 223 (3):136–46, 1970.

Edney, E. B., "Water Balance in Desert Anthropods," *Science* 156:1059–1066, 1967.

Gates, D. M., "Energy Exchange in the Biosphere," in F. E. Eckardt (ed.), *Functioning of Terrestrial Ecosystems at the Primary Production Level,* UNESCO, 1968, pp. 33–43.

———, "Climate and Stability," in *Diversity and Stability of Ecological Systems,* Brookhaven Symposia in Biol No. 22, 1969, pp. 115–127.

Goldberg, E. D., "The Oceans as a Chemical System," in M. N. Hill (ed.), *The Sea.* London: Interscience, 1963, Vol. 2, pp. 3–25.

Jackson, R. M., and Raw, F., *Life in Soil.* New York: St. Martin's, 1966.

Kuenzler, E. J., "Phosphorus Budget of a Mussel Population," *Limnology and Oceanography* 6(4):400–415, 1961.

Likens, G. E., F. H. Bormann, H. M. Johnson, and R. S. Pierce, "The Calcium, Magnesium, Potassium, and Sodium Budgets for a Small Forested Ecosystem," *Ecology,* 48:772–785, 1967.

Penman, H. L., *Vegetation and Hydrology,* Farnham Royal, England: Commonwealth Bureau of Soils, Technical Communication 53, 1963.

———, "The Water Cycle," *Scientific American* 223 (3):98–108, 1970.

Porter, W. P., and D. M. Gates, "Thermodynamic Equilibria of Animals with Environment," *Ecological Monographs,* 39:227–244, 1969.

Stewart, W. D. P., "Nitrogen-Fixing Plants," *Science* 158:1426–1432, 1967.

Thornthwaite, C. W., "An Approach to a Regional Classification of Climate," *Geographical Review* 38:55–94, 1958.

Whittaker, R. H., "Experiments with Radiophosphorus Tracer in Aquarium Microcosms," *Ecological Monographs*, 31:157–188, 1961.

Further Readings in Human Ecology

Aynsley, E., "How Air Pollution Alters Weather," *New Scientist* 44:66–67, 1969.

Bradley, C. C., "Human Water Needs and Water Use in America," *Science* 138:489–491, 1962.

Brown, L. R., "The World Outlook for Conventional Agriculture," *Science* 158:604–611, 1967.

Deevey, E. S., "Mineral Cycles," *Scientific American* 223 (3):149–58, 1970.

Edmonson, W. T., "Changes in Lake Washington Following an Increase in Nutrient Income," *Proc. Intl. Assn. Theoret, and Appl. Limnol.* 14:167, 1961.

Edmondson, W. T., "Water-Quality Management and Lake Eutrophication: The Lake Washington Case," in T. H. Campbell and R. D. Sylvester (eds.) *Water Resource Management and Public Policy*. Seattle: U. of Washington Press, 1968, pp. 139–178.

Feiss, J. W., "Minerals," *Scientific American* 209(3):128–136, 1963.

Hasler, A. D., "Cultural Eutrophication Is Reversible," *Bioscience* 19:425–431, 1969.

Lowry, A. D., "The Climate of Cities," *Scientific American* 217(2):15–23, 1967.

Revelle, R., "Water," *Scientific American* 209(3):92–108, 1963.

Sawyer, C. N., "Basic Concepts of Eutrophication," *J. Water Control Federation* 38:737–744, 1966.

Went, F. W., "Climate and Agriculture," *Scientific American*, 196(6):82–94, 1957.

Woodwell, G. M., "Radiation and the Patterns of Nature," *Science* 156:461–470, 1967.

Woodwell, G. M., and D. B. Wingate, "DDT Residues and Declining Reproduction in the Bermuda Petrel," *Science* 159:979–981,1968.

Review Questions

1. What is meant by a *sedimentary nutrient cycle?* Illustrate your answer by reference to one specific example.

2. What is meant by a *gaseous nutrient cycle?* Illustrate your answer by reference to one specific example.

3. Show how the hydrological cycle, the oxygen cycle, and the carbon cycle are interrelated and interdependent.

4. Discuss the influence of temperature on ecological processes.

5. What broad ecological effects does the nature and level of precipitation have in North America?

6. Describe what is meant by the term *evapotranspiration*. State what effects evapotranspiration can have on ecological processes.

7. What are climographs? Describe the applications of climographs in ecological studies.

8. Why is soil regarded as an exceedingly complex ecological factor?

9. State the various ecological effects of light. Illustrate these by actual examples, and mention instances where these effects are considerably reduced or even inoperative.

10. Describe the effects of fire on the nature and structure of ecosystems.

11. Briefly define *residence time* and *turnover rate*. What is the relationship between these two and by what alternative names may they be known? Discuss the ecological significance of these two concepts.

12. Discuss the ecological significance of microenvironmental studies.

3

Population Dynamics

In the previous chapters we have considered the general nature of ecosystems, and particular features of their structure and their function. We have examined the basic interactions between their biotic elements or communities and the principal chemical substances and environmental factors which comprise their abiotic element and together constitute their habitats. The remaining chapters of this text describe various community interactions and especially those which determine the size, origin, evolution, behavior, stability, and diversity of their component populations.

This third chapter deals with *population dynamics*, the study of the growth and structure of populations together with the factors that regulate their size and cause fluctuations in their density. This is an aspect of ecology that has long engaged the attention of population biologists and more especially of animal ecologists. Only comparatively recently has it been appreciated that plant populations exhibit similar dynamic relationships to those revealed by animals and microbes.

Perhaps because they were undertaken by workers who were essentially zoologists, botanists or microbiologists, studies of population dynamics on nonhuman species proceeded largely independently of those on human population dynamics. The latter were known as *demography*, and were undertaken by social scientists. Human populations, both by virtue of their great size and because of the unique records which are maintained regarding individual members, provide much valuable information on many aspects of population dynamics.

Demography

The term *demography* may in fact be applied to the study of the population dynamics of any organism. In its restricted meaning as the study of fluctuations in the size of human populations, it can be approached by constructing a hypothetical population and illustrating from this the several parameters used to describe samples taken from this population by various census procedures. Such a theoretical example is considerably simplified if, as shown in Table 3-1, it is postulated as

TABLE 3-1

Population growth in a human population, starting from a single couple and continuing initially by consanguineous matings. Each couple has six children all of whom survive to adulthood, and who die at age 60. One generation is taken as spanning 30 years. After three centuries and ten generations the population has grown to 157,464.

Generation	Number of Births	Number of Deaths	Total Population
1	6	—	8
2	18	2	24
3	54	6	72
4	162	18	216
5	486	54	648
6	1,458	162	1,944
7	4,374	486	5,832
8	13,122	1,458	17,496
9	39,366	4,374	52,458
10	118,098	13,122	157,464

initiated by a series of consanguineal matings. These are currently quite unusual in human societies, virtually all of whom have developed incest taboos. The Egyptian royal dynasty of whom Cleopatra was the penultimate product provides one of the uncommon instances in which brother-sister marriages were the custom.

There are two ways in which a population such as this theoretical one in Table 3-1 may acquire additional members. One is by *immigration,* the movement of members into the population from other populations, the other is by the occurrence of births or *natality.* Likewise there are two ways in which such a population may lose members, the one by *emigration,* the other by death of members or *mortality.* In this particular theoretical treatment it is assumed that neither immigration nor emigration occurs, and we will for the present examine only the standard demographic ways of expressing crude birth rate (natality) and crude death rate (mortality).

Birth Rates

For demographic purposes the birth rate is expressed as the total number of births in a year divided by the total population at midpoint of that year, that is on July 1, multiplied by 1,000. For our hypothetical example, we can greatly simplify the calculation by taking a mean figure. The birth rate at the midpoint of the last generation illustrated is obtained as follows:

Mean population on July 1 = 83,279

Mean total number of births in one year = 62,532

$$\text{Birth rate} = \frac{62,532}{83,279} \times 1,000$$

$$= 750$$

This theoretical birth rate is a magnitude higher than any which has actually been recorded for a human population. It can be compared with actual birth rates for contemporary populations as illustrated in Table 3-2. In the selection of statistics provided here, Dahomey has the high-

TABLE 3-2

Contemporary "crude" birth rate statistics from a selected series of countries. Figures are the number of live births per thousand head of population.

Dahomey	54	Chile	35
Niger	52	Uruguay	24
Cameroon	50	Japan	19
Nigeria	50	United States	17.6
Guinea	49	Italy	17.6
Ghana	47	United Kingdom	17.1
Madagascar	46	Belgium	14.8
Costa Rica	45	East Germany	14.3
Mexico	44	Sweden	14.3
Uganda	43	Luxembourg	14.2

Data from the 1970 World Population data sheet, reproduced with the permission of the Population Reference Bureau.

est figure, 54, followed closely by other West African nations such as Niger with 52. The lowest birth rate recorded for countries of any size are those of Sweden and East Germany with 14.3, only a little lower than those of the United Kingdom and the United States, 17.1 and 17.6 respectively.

The statistics shown in Table 3-2 suggest that birth rates seem mostly to be either in the 40's and 50's or between 14 and the low 20's. This is not a random distribution of numbers as the graph in Fig. 3-1 shows. It is attributable to the circumstance that some nations have adopted population-control methods on an extensive scale, and brought their birth rates down to below 20, while others as yet have not, in which case the birth rates stay up in the 40's and 50's. Demographers refer to this cultural adjustment by a society to a lower birth rate as the *demographic transition*.

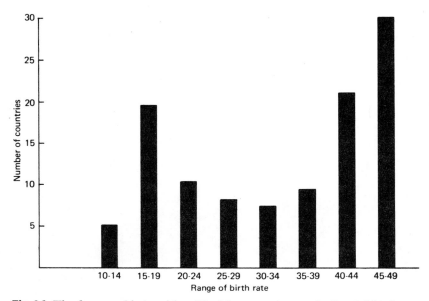

Fig. 3-1. The demographic transition. The histogram presents the "crude" birth rate statistics for 116 countries, obtained from the 1970 World Population Data Sheet, prepared by the Population Reference Bureau. Birth rates are expressed as the annual mean number per 1000 of population. There are two peaks, one in the 15-19 class, another in the 45-49 class. The first represents countries which have undergone the demographic transition. In the second class, birth rates have not yet been adjusted downwards by cultural procedures. Where mortality rates have been lowered, countries in this second class will now be experiencing a population explosion.

Mortality Rates

The death rate or mortality is calculated in a very similar way to the birth rate, from the total number of deaths which occur during the year divided by the total population as at the midpoint of that year and muliplied by a thousand. In the theoretical example given in Table 3-1 the total number of deaths in the final year is

$$\text{Mean population July 1} = 83,279$$
$$\text{Mean total number of deaths in one year} = 6,948$$
$$\text{death rate} = \frac{6,948}{83,279} \times 1,000$$
$$= 83$$

Death-rate statistics are also provided by the Population Reference Bureau; a current selection of these is listed in Table 3-3. West Africa again heads the list, with Guinea and Cameroons standing at 26. Hong

TABLE 3-3

Contemporary "crude" mortality statistics from a selected series of countries. Figures are the number of deaths per thousand of population.

Guinea	26	Luxembourg	12.3
Cameroons	26	United Kingdom	11.9
Dahomey	26	Chile	11
Nigeria	25	United States	9.6
Niger	25	Netherlands	8.2
Uganda	18	Ceylon	8
Pakistan	18	Costa Rica	8
India	17	Japan	7
East Germany	14.3	Singapore	6
Belgium	12.8	Hong Kong	5

Data from the 1970 World Population data sheet, reproduced with permission of the Population Reference Bureau.

Kong at 5 and Singapore at 6 have the lowest rates recorded. The reasons why the death rate in these areas is much lower than in advanced industrial societies such as the United States, the United Kingdom, and Belgium, which have death rates of 9.6, 11.9, and 12.8 respectively, are due to differences in population structure which will be examined shortly.

The remarkable reduction in death rates in such countries as India, Pakistan, Chile, and Ceylon, which is apparent from Table 3-3, occurred only as recently as the 1950's. It was almost 100 years later than the parallel reduction which occurred earlier in the industrialized societies. It is believed due to the large-scale introduction of two general medical procedures which are highly effective in preventing wide-scale epidemics of previously fatal diseases. One was the use of DDT and (later) other pesticides to control insect vectors of diseases such as malaria and typhus. The extensive use of pesticides for medical and other purposes has unfortunately had effects other than those intended, as is discussed in the volume on pesticides in this series. The discovery of drugs with antibiotic effects, first the sulfa series, followed quickly by penicillin and other antibiotics, coupled with the introduction of prophylactic serums, provided protection against the fatal effects of diseases like cholera, dysentery, infantile paralysis, pneumonia, and yellow fever. This was the second major medical innovation which drastically reduced mortality in underdeveloped nations after World War II.

Rate of Natural Increase

The rate of natural increase is the usual demographic statistic used to express population growth. It is obtained by subtracting the mortality rate from the birth rate. Thus in the hypothetical example in Table 3-1 the mortality rate of 8.3 deducted from the birth rate of 75 gives a figure of

66.7. Instead of continuing to express this as a value per thousand persons in the population, it is more usual to convert it to a percentage figure, and the rate of natural increase therefore becomes 6.6 percent.

Table 3-4 lists a range of rates of natural increase selected from the

Percent rate of natural increase statistics from a selected series of countries. These do not always coincide with the figures calculated from the crude birth and death rates. The Bureau has sometimes adjusted them to allow for migratory movements, or what it believes to be a misleading figure for the birth or death rate.

Costa Rica	3.8	Ceylon	2.4
Mexico	3.4	Guinea	2.3
Pakistan	3.3	Cameroon	2.2
Niger	2.9	Luxembourg	1.2
Madagascar	2.7	Uruguay	1.2
India	2.6	Japan	1.1
Nigeria	2.6	United States	1.0
Dahomey	2.6	United Kingdom	0.5
Hong Kong	2.5	Belgium	0.4
Singapore	2.4	East Germany	0.3

Data from the 1970 World Population data sheet, reproduced with permission of the Population Reference Bureau.

statistics for a number of nations. East Germany which shares the lowest birth rate of a country of any size, with a fairly low death rate, has a natural increase of 0.3 which is a close approximation to a stable population—one that is neither increasing nor decreasing. West African countries like Dahomey and Niger which had the highest birth rates do not have the highest rates of natural increase because their mortality rates are still quite high. Some Latin American countries which have high birth rates have also in general the highest rates of natural increase because of relatively low death rates. Costa Rica leads with a rate of 3.8 percent. One statistic, that for Kuwait, is not included in the table because it is a tiny population and migratory movements cause it to be very misleading; Kuwait is actually recorded as having a birth rate of 47 and a death rate of 6, with a natural increase of 8.3 percent, just over twice that of Costa Rica.

In order to provide a more readily comprehensible statistic for population increase, demographers sometimes convert the percentage rate of population increase to *population doubling time*. This is the length of time it would take for the size of the population to double if the existing rate of natural increase continued unchanged. Table 3-5 shows the equivalence in doubling time for integers of rates of natural increase.

The predictable doubling times for a selection of countries based on

TABLE 3-5

Population doubling times, to the nearest integer for a given series of rates of natural increase. As may be seen from Table 3-4, no national population has yet achieved a rate of natural increase as low as 0.1 percent. A few centuries only will therefore see all human populations at least doubled in number. Unfortunately many populations still cluster around the lower portion of this table. Several centuries at these rates would see these populations reach entirely unsupportable levels.

Percent Rate of Natural Increase	Population Doubling Time in Years
0.1	500
0.5	150
1.0	70
2.0	35
3.0	24
4.0	17
5.0	12
6.0	9

current rates of natural increase are provided in table 3-6. East Germany, which as already noted has almost stabilized its population, will take 233 years to double its population size at present rates of natural increase.

TABLE 3-6

Population doubling time for a selected series of countries, in years.

Costa Rica	19	Ceylon	29
Mexico	21	Guinea	31
Pakistan	21	Cameroon	32
Niger	24	Luxembourg	58
Madagascar	26	Uruguay	58
India	27	Japan	63
Nigeria	27	United States	70
Dahomey	27	United Kingdom	140
Hong Kong	28	Belgium	175
Singapore	29	East Germany	233

Data from the 1970 World Population data sheet, reproduced with the permission of the Population Reference Bureau.

Costa Rica, with the highest rate of natural increase, will take only 19 years to double its population. This is almost quadrupling the size of the population in every human generation. In the exceptional case of Kuwait the fantastic doubling time of nine years is indicated, allowing for an eightfold increase in every generation.

The demographic characteristics of human populations as illustrated from these very basic statistics may be better understood by reference to the work of population ecologists on other populations, and especially on other animal populations.

Growth Characteristics of Populations

When a culture is prepared for any microbial population, such as that of a yeast, the population increase after certain intervals of time may be estimated by withdrawing samples of the culture and counting the number of microbes present in a standard volume of the sample. When the calculated size of the population at various time intervals is plotted as in Fig. 3-2, a sigmoid type of growth curve results. This curve

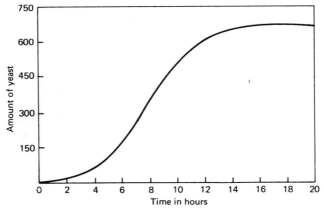

Fig. 3-2. Growth of yeast in culture illustrating the sigmoid type of population increase. After the initial "lapse time" the growth rate becomes exponential. This rate of growth is slowed as the asymptote is reached until it is reduced to zero. The carrying capacity of this culture is set by the amount of nutrients initially available in the medium. This type of growth curve is said to be "logistic"—see text for a further explanation.

has three portions—an initial one in which population size increases slowly during what is known as the *lapse time,* followed by a period of extremely rapid exponential growth often called the *logarithmic phase,* and culminating in a final phase during which population growth gradually falls away until it eventually ceases altogether. The lack of further increase is due to such factors as the accumulation of toxic products in the culture, or the exhaustion of some essential food resource.

This sigmoid type of population growth as expressed by a unicellular organism such as yeast is not universally typical of the population growth either of all unicellular or of all free living multicellular organisms, as illustrated in Fig. 3-3. In this second example of population growth the early phases are very similar. Initially there is a slow build-

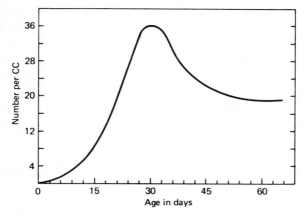

Fig. 3-3. Growth of water fleas *(Daphnia)* **in culture** illustrating the J-shaped type of population growth pattern. After the initial "lapse time" the rate of natural increase becomes exponential as in Fig. 3-2. It continues with this form of growth until population density suddenly and markedly reduces the number of young animals added to the population. This causes a population "crash," from which there may be fluctuating recovery.

up, followed by a more rapid or logarithmic phase giving the so-called J-shaped curve. However instead of gradually falling off and reaching a steady state equilibrium at the asymptote, the curve drops back dramatically as population growth crashes, and is very drastically reduced to a much lower level. The crash is again the result of the intervention of some limiting substance or factor, such as the exhaustion of food resources or overcrowding of space. We have only just begun to relate the curves we can construct from demographic statistics to these characteristic types of curves representing nonhuman population growth.

Two ecological terms have been introduced to describe the factors responsible for the basic similarities of both these initial types of population growth. The first is *biotic potential,* the continuing reproduction of the organisms at a maximum rate under favorable conditions. The second is *environmental resistance,* which causes the population growth to reach a steady state or decline instead of continuing at the maximum rate. Environmental resistance is thus the totality of limiting substances and factors which prevent the continuing attainment of the biotic potential.

The Intrinsic Rate of Natural Increase

The biotic potential is usually expressed by a factor representing the intrinsic rate of natural increase r. This is the instantaneous rate of increase per head of a population under particular environmental conditions, when the effects of changes in population size can be ignored. It is contained in the following expression:

$$rN = \frac{dN}{dt} \tag{1}$$

this is a differential equation which becomes when integrated:

$$N_t = N_o e^{rt} \tag{2}$$

where N_t = number of individuals at time t
N_o = number of individuals at time zero
e = base of natural logarithms (2.718)
t = elapsed time

Using natural logarithms this converts to

$$\log_e N = \log_e N_o + rt \tag{3}$$

plotting population growth against time using this equation provides a straight-line relationship (Fig. 3-4).

Environmental Resistance

Expression (1) may be modified to provide for the operation of the environmental-resistance factor which is usually indicated by K, giving the expression:

$$\frac{dn}{dt} = rN \frac{(K\text{-}N)}{K} \tag{4}$$

Whereas expression (1) approximates to the J-curve of exponential population growth (Fig. 3-3), expression (4) corresponds to the sigmoid form to which many populations at least approximately conform (Fig. 3-2). This second type of population growth is sometimes said to be following a *logistic curve*, but it must be noted that population-growth curves are not actually logistic, nor is the logistic curve always a good fit with a sigmoid curve for growth. Nevertheless the sigmoid growth curve is a useful concept. It provides a single form of general curve of population growth illustrating variation in the intrinsic rate of increase arising from interaction between the biotic potential and environmental resistance.

The factor K for environmental resistance is more generally incorporated within the concept of the *carrying capacity*. This represents the population size at which some limiting substance or factor or a combination of these prevents further population increase. It is an extremely important concept when considering optimum population size.

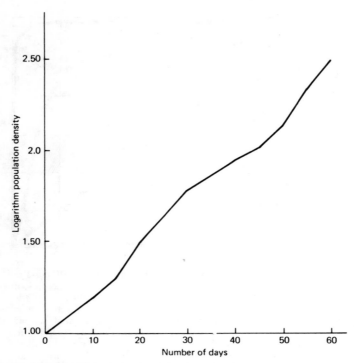

Fig. 3-4. Logarithmic expression of population growth. The same data have been used to construct this graph as those which provided the J-shaped population growth curve in Fig. 3-3. The size of the population has however been converted to logarithmic terms. The result is a straight line relationship, demonstrating that growth of the *Daphnia* population during this phase is logarithmic.

Survivorship

Whether population growth is recorded in this way as the intrinsic rate of natural increase, or whether it is determined demographically as the percentage rate of natural increase, it is a measure of the difference between the natality and mortality in a population. Whereas natality, the addition of members to the population by birth, is a relatively simple factor to express, mortality may be conceived in several different ways. Because ecologists are interested primarily in the organisms that *survive* rather than those that die, mortality is commonly expressed in the form of what is known as a *survivorship curve*. Several types of survivorship curves are illustrated in Fig. 3-5.

These curves indicate the number of survivors out of an initial population of a given size after a specified time interval, assuming that all the individuals in the population at the beginning of the observations

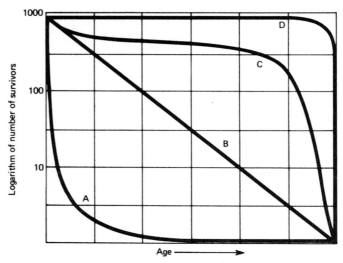

Fig. 3-5. Survivorship curves of four different types. All populations are of similar-aged individuals, and the graphs represent the numbers surviving after given intervals of time. A is the curve for an oyster population; B, a population of fledged birds; C, a population of mountain sheep; D, a population of starved fruit flies. The figures on the vertical axis, which is logarithmic, show the number of individuals surviving from an original population of 1000. The age-scale intervals on the horizontal axis vary with the life span of the particular species. (Data from various sources.)

were of similar age. It also assumes that all members of the population orginally had the same capacity for survival and that environmental effects are ignored.

In the first form of survivorship curve, in this case representing a population of equal-aged starved fruit flies under laboratory conditions, individual flies survived for almost equal lengths of time. The curve therefore shows no diminution of survivors until very close to the point when mortality is universal. This gives a right-angle form of curve to which, as can be seen from the figure and from Fig. 3-6 the survivorship curve for the human population of an industrialized society is also beginning to conform. This is because mortality losses among human infants and juveniles have been very greatly reduced in such societies, as have losses among senior citizens in their fifties and early sixties. As nevertheless the actual length of duration of human life has not been substantially increased, this is tending to concentrate the ages at which people die into their late sixties, seventies, and early eighties.

Almost the opposite form of graph is shown by many wild invertebrate populations, such as that illustrated for oysters, and for most

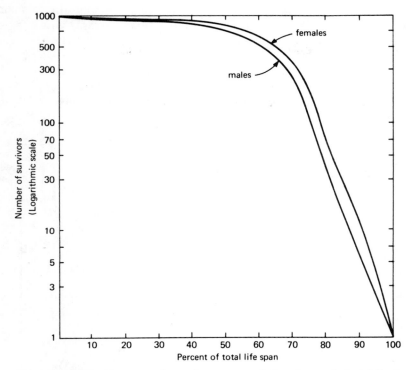

Fig. 3-6. Survivorship curve for a human population in an industrialized country. Lowering infant and juvenile mortalities produce a pattern which with improving health services conforms increasingly to that of the starved fruit-fly population in Fig. 3-5. Mortality is light until individuals approach the average life-span of approximately 70 years. Female mortality is slightly lower than male at most ages.

plant populations (Fig. 3-5). In this instance there is a high embryonic or infant and juvenile mortality, while the members of the population who survive these early phases subsequently enjoy an extremely low mortality.

A straight-line survivorship relationship, as is illustrated in Fig. 3-5 for a fledged-bird population of equal aged individuals, represents a form of survivorship curve in which mortality occurs at a constant rate throughout the period of life span of individuals in the population.

Within a given form of such survivorship curves there are differences arising from a number of reasons. One of the most consistent of these in heterosexual species is the difference in the form of the survivorship curve of males as compared to females of the same species. This sexual difference in human populations is illustrated in Fig. 3-6. As might be anticipated, it is these differences that cause actuarians to propose differentials between the premiums for life insurance and annuities, females being required to pay less for life insurance, more for annuities.

Other differences in the form of survivorship curves which may occur result from variations between different races of the same species, or different mutants, as has been extensively illustrated for the fruit fly *Drosophila* (Fig. 3-7). The form of the survivorship curve is also partially determined by the initial density of the population, which if higher than optimal may reduce individual chances of survival.

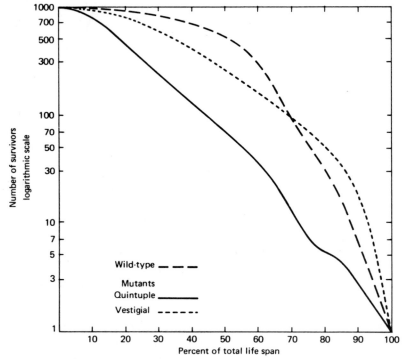

Fig. 3-7. Survivorship curves for mutant forms of fruitfly (Drosophila). The wild type has a longer mean life span (41 days) than the vestigial wing mutant (17 days) and the quintuple mutant (8 days). The life span figures on the horizontal axis are therefore expressed as a percentage rather than an actual value. (Data from several sources.)

Population Structure

The even-aged populations on which such survivorship curves have been based must clearly change in structure with time. They start initially as young populations and mature together to old ones. If, however, instead of starting with individuals of the same age a stable population is selected in the first instance, it will be composed of individuals of various ages in proportions which tend to remain constant. Bodenheimer in 1958 in an extensive review of the question of age structure in populations, suggested a basic subdivision of such a population into three categories, *preproductive, reproductive* and *postreproductive.*

If the survivorship curve prepared for such a stable population were of the form illustrated for an oyster population (Fig. 3-5), we would expect a pyramidal distribution of age classes in the population as illustrated in Fig. 3-8, with a very large prereproductive phase. The starved fruit-fly population (Fig. 3-5) by contrast would give an age structure diagram in which these basic age classes are almost uniform (Fig. 3-8). The form assumed by human populations is illustrated in the diagram in Fig. 3-8 for *young, stable,* and *declining* populations respectively.

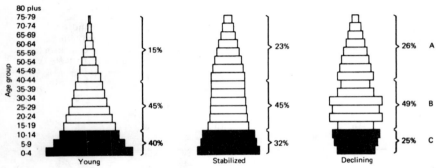

Fig. 3-8. Comparison of young, stabilized, and declining human populations. Although the size of the reproductive segment (B) is not widely contrasted in these three conditions, those of the prereproductive (C) and postreproductive (A) segments are. Any factor that markedly reduces the mortality rate in the prereproductive element of a young population will therefore produce a population explosion a few years later unless the fertility rate is reduced.

Additional Parameters of Human Populations

It can now be seen how the demographic statistics of birth and mortality rates, and the rate of natural increase for human populations, described earlier in this chapter, involve a number of inaccuracies. In a young population, mortality rates will be much lower than in a declining one, for there will be a higher proportion of younger people, whose mortality rate is lower. This is the reason why populations such as that of Hong Kong illustrated in Table 3-3, attain as low a mortality rate as five. It also explains why it is not sufficient in such populations, if the objective is the attainment of zero population growth, to restrict the number of offspring per married couple to two. Until the population age structure assumes a stable form, zero population growth may be attained only by keeping the number of offspring lower than two as is illustrated in Table 3-7.

TABLE 3-7

Fecundity and stable populations. By making a number of assumptions regarding the structure of a human population it is possible to derive the table below. This shows the fecundity rate which will produce zero population growth for given levels of mortality and birth rates. This table illustrates the point that for societies which have achieved a mortality rate lower than 15 the fecundity rate must be less than 2.0. In other words, a stable population in existing industrialized countries, and in areas like Hong Kong and Singapore with a high percentage of young people, can only be attained by keeping the average number of children per female to *less than two*. Indeed, in the last two instances it has to be *less than one*. In the light of such figures proposals like that for an "International childless year" appear less unrealistic.

When Mortality Rate Reaches This Figure	Birth Rate to Maintain Zero Population Growth Must Be	Fecundity (Number of Offspring per Female) to Provide This Birth Rate Is
45	45	6.7
35	35	5.2
25	25	3.7
15	15	2.2
10	10	1.5
5	5	0.7

Age structure also affects the birth rate in a rather similar manner, for the number of births per thousand of the population will be much higher if it contains a larger than usual number of females, or if it contains an undue proportion of young people. For this reason demographers frequently use instead the *fertility rate,* which is the number of births per thousand women in the population between the ages of 15 and 44. The fertility rate in the United States is illustrated in Fig. 3-9.

This still leaves us without information as to the actual amount of reproduction by individual female members of the population. In human populations this is commonly expressed by what is generally known as *fecundity rate,* the average number of live offspring born to an individual married woman (Table 3-8).

Whether we are dealing with human populations, or with those of plants, microbes, or other animals, and whatever parameters we choose to measure such features as natality, mortality, and natural increase, in ecosystems that have a steady-state balance, population growth has to approximate that of a stable population. The great majority of populations encountered in contemporary communities are therefore stable.

TABLE 3-8

Fecundity rates—variations in the number of live births per female in various different populations. These fecundity figures vary very considerably according to whether the population has undergone the "demographic transition." Underdeveloped countries have invariably high fecundity rates, but not necessarily high rates of natural increase. Whether they do so or not will depend on their mortality rate.

	Live Births per Female (to nearest integer)
Eskimos (hunters)	3
North American Indians (hunters)	3-4
Australian aborigines (hunter-gatherers)	5
Central Africa (agriculturalists)	ca. 6
India (agriculturalists)	6-8
U.S. (mid-twentieth century-industrialists)	3
U.S. (beginning nineteenth century—preindustrial stage)	7

Considerable attention has been given to determining the factors responsible for regulation of population numbers so as to achieve this steady state. A number of the volumes in this series deal further with this question of structure, stability and the steady state.

Regulation of Population Numbers

Animal ecologists have devoted most attention to the question of population regulation, which has attracted less emphasis in respect of plant and microbial populations. Two schools of thought have developed,

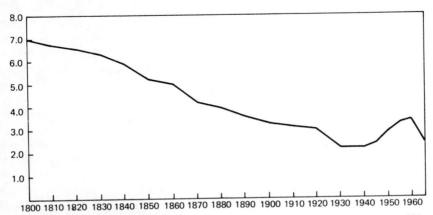

Fig. 3-9. Fertility rate for the U.S. population—mean rate for whites 1800 to 1970. The rate has steadily declined from the beginning of the nineteenth century to the depression years of the 1930-40 decade. The subsequent slight upward trend since then was reversed in the last decade, and the 1970 census will probably reveal the lowest fertility rate yet recorded. (Data from several sources.)

the one regarding populations as regulated by limiting substances or factors of either the biotic or the abiotic system according to their density, the so-called *density dependent* factors. The other school envisages that abiotic features operate to regulate population size independently of density, thus constituting what has been called the *density independent* factors.

For example, an experimental culture of fruit flies in a bottle will have its population size determined generally by the amount of food originally present in the culture medium. The number of fruit flies increases at first until there is insufficient food to meet the minimum requirement of every individual. The mortality then rises, and the rate of natural increase—that is, *population growth*—decreases, even if the birth rate remains steady. Food under these circumstances is considered to be a density-dependent factor regulating the size of the experimental *Drosophila* population.

A laboratory population of mice can be provided with an abundance of food so that it never becomes a limiting factor. As population growth continues there will be insufficient *space* in the cages in which the experiment is being conducted, and space becomes a limiting factor which is density-dependent.

In other laboratory experiments we could investigate the effect of temperature on population growth in fruit-fly cultures. As temperatures approached the upper or the lower tolerance limits we would find that population growth was reduced. Temperature is then functioning as a density independent factor, for population growth is reduced at the lower or upper limits of tolerance whatever the density fruit flies have attained in the culture.

This density-dependent and density-independent argument is now diminishing, and it appears likely to be agreed eventually that basically all population growth is density-dependent (Orians, 1962). Before becoming so however, it may pass through a phase in which it is regulated by density-independent factors.

Fluctuations in Numbers

Population density may be defined as the number of individuals per square unit of two-dimensional space, or per cube of three-dimensional space, whichever is the more appropriate. Ideas as to the form that fluctuations of numbers in animal populations and of the feedback mechanisms that regulate them may take are illustrated in three classic examples. The first is the case of the Kaibab deer which overexploited their habitat when predators were removed and overshot their food supply, adopting a J-shaped form of population growth followed by a subsequent population crash. The second example, made famous by Andrewartha and Birch, resulted from their study of the effect of meteorological factors on the intrinsic rate of increase of thrips populations (Davidson and Andrewartha, 1958). The third was the effect of overcrowding on a population of white rats under experimental conditions in the laboratory described by Calhoun (1962).

Population Explosion in Kaibab Deer

The Kaibab plateau, situated on the north rim of the Grand Canyon in Arizona, at the beginning of this century carried a herd of deer estimated at about 4,000 head. Over a period starting in 1907 and lasting for about 30 years, deliberate steps were taken to remove the main predators of deer from this area, namely wolves, mountain lion, and coyotes. The effects of this removal of the predators on the population growth of the deer population is shown in Fig. 3-10.

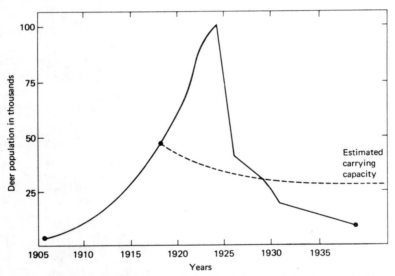

Fig. 3-10. Population fluctuation in the Kaibab deer, following eradication of the main predators, wolves, coyotes and cougars. See text for further explanation. (Data from several sources.)

By 1924 the numbers of deer are estimated to have risen to 100,000 from the original 4,000. As may be seen from the illustration, population growth followed a J-shaped curve. At this 1924 peak population density, the numbers of deer had far overshot the estimated carrying capacity for deer of the Kaibab plateau, which was in the region of about

30,000. The resultant overgrazing of the area caused damage to the range and a considerable reduction in its carrying capacity. In the years 1924, 1925, and 1926 it is estimated that some 60 percent of the deer population died from starvation and disease. By 1939 about 10,000 deer were existing on the range, considerably more than the original population of 4,000 because of the virtually total removal of predators. This was also a substantially lower figure than the estimated 30,000 carrying capacity of the area before it was damaged by overgrazing.

Seasonal Thrips Populations

The results of one of a number of experiments conducted by Andrewartha and Birch are illustrated in Fig. 3-11 which show the

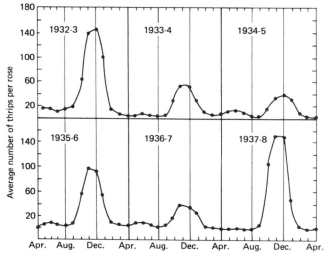

Fig. 3-11. Population fluctuations in rose thrips. Both seasonal changes in population density and annual variations in maximum population size are density independent according to Andrewartha and Birch. The rate of increase of such thrips populations they believed was determined by local microenvironmental factors and the proximity of existing thrips populations. (Redrawn: based on Andrewartha and Birch, 1954.)

seasonal change in a population of adult thrips observed on rose bushes in Australia. Variations in population density of the thrips not only occur seasonally, but this seasonal variation varies in extent from year to year. Andrewartha and Birch represented that a knowledge of the local meteorology, the rate of mobility of the thrips, and the distance between the infected rose bushes and the various thrips populations from which they became infested, enabled a forecast to be made of the extent of these seasonal and annual fluctuations. Not all workers subsequently agreed with this interpretation of their observations (Smith, 1963).

Effects of Overcrowding

In an often quoted experiment by Calhoun (1962), laboratory rats were provided with adequate water, food, and nesting materials, and housed in cages constructed as shown in Fig. 3-12. The rat population increased

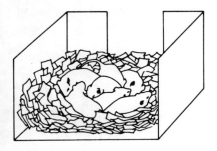 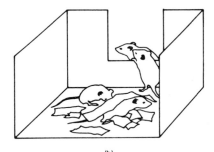

(a)   (b)

Fig. 3-12. The effect of population density on the social behavior and organization of a laboratory rat colony. The litter shown in (a) was born to a mother still undisturbed by the level of population density. A female disturbed by the level of population density failed to complete an adequate nest (b). The young left the litter precociously, fewer than five percent survived. (Redrawn: based on Calhoun, 1962.)

until it reached a steady state as expressed in the asymptote of a sigmoid type of population growth. This ultimate regulation of the intrinsic rate of increase appeared to have resulted from interaction between overcrowding and the functioning of the endocrine glands which determine the rats' function and behavior. The animals in this final phase exhibited various forms of perversion and could not reproduce normally, sometimes failing to copulate, or neglecting to care for the young. Some embryos were reabsorbed, others were aborted; young were often cannibalized. Some females apparently became nonfertile. The sum total of these effects was to reduce population growth until it reached a steady state in which these various behavioral and physiological reactions occurred at a sufficient intensity to maintain a balance between birth rate and mortality rate.

In this instance spacing appears to have acted as a feedback mechanism through its effect on the endocrine glands. A more commonly encountered population regulation mechanism reflecting spatial interactions is the phenomenon known as *territoriality*.

Territoriality

Territorial concepts of population regulation were originally developed from behavioral studies on birds. The phenomenon has subsequently been shown to exist in a wide range of animals from large mammals like bears to small ones such as deer mice. Groups including lizards, fishes, and some social insects are also involved (Fig. 3-13).

An example of a bird territory is shown in this figure. Within its own territory the male of the bird species concerned will drive away any other male of the same species which ventures into the territory. This is defined by intangible boundaries which we are not able presently to recognize, but can outline by charting the movements of individual males. One of the functions of bird song appears to be the establishment of territorial rights in this individual area.

Very often the size of the territory is related to the amount of food required to raise a brood of fledglings. In a bad year when food is scarce, the territories tend to be larger, and the number of unmated birds left without territory greater. The more limited food is thus spread over a less dense population. In a favorable year, territory sizes are smaller, there are more mating pairs and more broods are raised, and there are fewer unmated birds.

Social primates such as baboons have what is known as a *nucleated territory*. This is a type which is not defended at its boundaries, as in the case of a bird territory, but where antagonism between invading social groups become greater the closer they penetrate into the heart of the territory. Although again there are no perceptible boundaries marking the nucleated territory, other social groups of the same species patently become nervous approaching the boundaries of another group's territory.

Birdsell (1953) showed that hunting-gathering human societies occupied nucleated territories. In work which he performed on the dispersal patterns of Australian aborigine groups, Birdsell was able to demonstrate that the size of these nucleated territories was inversely related to the rainfall of the area.

Regulation of Human Populations

We have lately turned to more intensive exploration of the possible existence of regulatory mechanisms in our own species. Two contrary schools of thought exist. One is based on the work of Malthus (1790), which supposes that human populations are always maintained at a *maximum* determined by the operation of density-dependent limiting

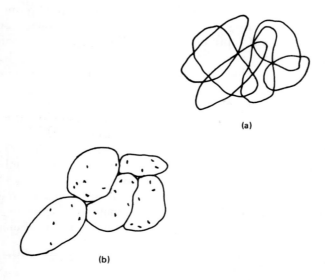

(a)

(b)

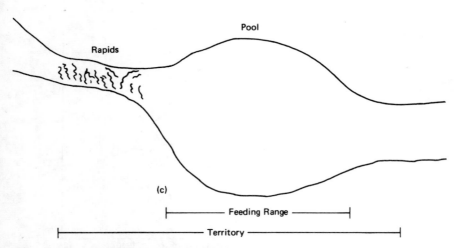

(c)

Fig. 3-13. **Territorial behavior in various animals.** Diagrammatic outlines of the territories of several individuals in a species of tortoise (a), bird (b), of a single male fish (c). In (b) the positions of roosting perches within each territory are indicated by dots. In (c) the feeding range in the pool is more restricted than the portion of the habitat defended against the intrusion of other male fish of the same species. Tortoise males do not defend a territory, hence the overlap shown in (a). (Redrawn from several sources.)

factors of war, famine, and pestilence. The other school founded by Carr-Saunders in 1926 supposes that there is an *optimum* human population for each ecosystem. This optimum is maintained by the operation of various limiting but equally density-dependent *cultural* factors. These regulate population size so that it never escapes to the maximum, where more extreme regulatory limiting factors would intervene.

Some human societies have patently escaped from the regulatory mechanisms of such cultural factors as might maintain the population at an optimum density. Many of the environmental crises with which we are now confronted arise from the seeming inevitability that war, pestilence, famine, or some combination of these are the only regulatory mechanisms remaining to check continued population growth within some national boundaries. In order to reestablish cultural regulatory mechanisms, nations must go through what sociologists call the *demographic transition*. As already noted this refers to the circumstances that

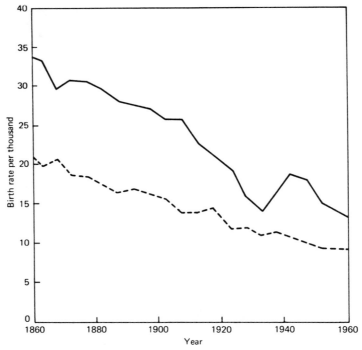

Fig. 3-14. The Demographic transition illustrated for an industrial nation over the last century. This shows the birth rate (solid line) adjusts to the declining mortality rate (dotted line). These data for Sweden display the essentially unexplained reversal of the downward trend in the birth rate in the 1930's and early 1940's which affected every Western nation. This graph should be related to the histogram in Fig. 3-1.

when mortality rates become drastically reduced through the introduction of modern medicine and hygiene, the birth rate also has to be lowered by the adoption of population-control techniques which restrict the number of births and balance the reduced death rate to achieve zero population growth.

The industrialized societies of the West achieved this demographic transition through the end of the nineteenth century and the first half of the twentieth century by lowering their birth rates parallel with their mortality rates (Fig. 3-14). As was seen earlier in the chapter, some of these nations, like East Germany and Belgium, have now done this so successfully that they have achieved what is virtually zero population growth. No nation outside the western European powers, and the populations descended from them, have completed this demographic transition except Japan, which until several years ago appeared also to have achieved a stabilized population by this means. As will be referred to repeatedly in the volumes in this series, many careful studies suggest that the world population is already far too high for it ever to be possible even at present densities to achieve the same technological standard of living as presently prevails in the United States. As the figures in Table 3-6 showing various population doubling times illustrate, the momentum of population increase cannot now be slowed until densities for the world population far higher than present are reached. We are therefore faced with only two possibilities. The one is to recognize that no further technological improvements will be possible for any group, whatever its present level, thereby condemning a large portion of the world to a minimal standard of living at bare subsistence levels. The other is to aquiesce in action drastically to reduce population densities in the more overcrowded areas, until these reach a calculated optimum population for which there are sufficient resources and space. How this may be achieved is reviewed by Berelson (1969).

Bibliography

References

Calhoun, J. B., "Population Density and Social Pathology," *Scientific American* 206(2):139–148, 1962.

Chitty, D. H., "Population Processes in the Vole and their Relevance to General Theory," *Canadian Journal of Zoology* 38:99–113, 1960.

Frank, P. W., "Prediction of Population Growth Form in *Daphnia pulex* Cultures," *American Naturalist* 94:357–372, 1960.

Orians, G. H., "Natural Selection and Ecological Theory," *American Naturalist* 96:257–263, 1962.

Palmead, I. G., "Competition in Experimental Populations of Weeds with Emphasis on the Regulation of Population Size," *Ecology* 49:26–34, 1968.

Park, T., "Beetles, Competition, and Populations," *Science* 138:1369–1375, 1962.

Preston, F. W., "Diversity and Stability in the Biological World," in *Diversity and Stability in Ecological Systems,* Brookhaven Symposium in Biology No. 22, 1969, pp. 1–12.

Rose, S. M., "Feedback Mechanism of Growth Control in Tadpoles," *Ecology* 41:188–199, 1960.

Slobodkin, L. B., *Growth and Regulation of Animal Populations.* New York: Holt, 1961.

Solomon, M. E., *Population Dynamics.* New York: St. Martin's, 1969.

Smith, F. E., "Density Dependence," *Ecology* 44:220, 1963.

————, "Population Dynamics in *Daphnia magna* and a New Model for Population Growth," *Ecology* 44:651–663, 1963.

Whittington, W. L., and O'Brien, T. A., "A Comparison of Yields from Plots Sown with a Single Species or a Mixture of Grass Species." *Journal of Applied Ecology* 5:209–213, 1968.

Wynne-Edwards, V. C., "Population Control in Animals," *Scientific American* 211(6):68–74, 1964.

Further Readings in Human Ecology

Berelson, B., "Beyond Family Planning," *Science* 164:533–543, 1969.

Bradley, C. C., "Human Water Needs and Water Use In America," *Science* 138:489–491, 1962.

Brown, L. R., "Human Food Production as a Process in the Biosphere," *Scientific American* 223(3):161–170, 1970.

Cook, R. C., "California after 19 Million What?," *Population Bulletin* 22:29–57, 1966.

Davis, K., "Population," *Scientific American* 209(3):62–71, 1963.

Dorn, H. F., "World Population Growth: an International Dilemma," *Science* 135:283–290, 1962.

Doxiadis, C. A., "Man's Movement and his City," *Science,* 162:326–334, 1968.

Feiss, J. W., "Minerals," *Scientific American* 209(3):128–136, 1963.

Fredericksen, H., "Feedbacks in Economic and Demographic Transition," *Science* 166:837–847, 1969.

Heer, D. M., "Economic Development and the Fertility Transition," *Daedalus* 97:447–462, 1968.

Howard, W. E., "The Population Crisis Is Here Now," *Bioscience* 19:779–784, 1969.

Lowry, W. T., "The Climate of Cities," *Scientific American* 217(2):15–23, 1967.

Malthus, T. B., *First Essay on Population.* 1798 (Reprinted, London: Macmillan, 1926).

Markert, C. L., "Biological Limits on Population Growth," *Bioscience* 16:859–862, 1966.

McElroy, W. D., "Biomedical Aspects of Population Control," *Bioscience* 19:19–23, 1969.

Pirie, N. W., "Orthodox and Unorthodox Methods of Meeting World Food Needs," *Scientific American* 216(2):27–35, 1967.

Population Reference Bureau, 1970 World Population Data Sheet, April 1970.

Revelle, R., "Water," *Scientific American* 209(3):92–108, 1963.

Robinson, H. F., "Dimensions of the World Food Crisis," *Bioscience* 19:24–28, 1969.

Scrimshaw, N. S., "Food," *Scientific American* 209(3):72–80, 1963.

Sears, P. B., "The Inexorable Problem of Space," *Science* 127:9–16, 1958.

Spilhaus, A., "The Experimental City," *Science* 159:710–715, 1968.

Young, G., "Drylands and Desalted Water," *Science* 167:339–343, 1970.

Review Questions

1. Define the terms *natality, mortality, natural increase*. Discuss the relationships these terms indicate, and the significance of these in relation to population growth.

2. Describe what is meant by *fertility rate, fecundity, gross reproductive rate*. What are the advantages and disadvantages of using each one of these statistics as an indication of natality?

3. Discuss what is meant by the term *population structure*. What are the characteristics of the population structure of (a) a young population, (b) a stable population, (c) a declining population?

4. What are the classically defined differences between density dependent and density independent factors? What significance is currently attributed to these differences in theory and in practice?

5. What types of feedback mechanisms in natural ecosystems influence the rate of increase of their component populations?

6. What is a survivorship curve? Describe the major types of variations encountered in the form of survivorship curves. Provide named examples of each type you mention.

7. Describe the condition commonly identified by the term "zero population growth." Explain how it is possible in a sexual species to have a population with a gross reproductive rate of 1.0 yet find a rate of natural increase which is either positive or negative.

8. What do you understand by the term *secondary sex ratio*? Provide and discuss examples of variation in this ratio.

9. State what is meant by the *demographic transition*. Discuss the reasons why nations fall into two major groups in respect to this feature.

4

Population Evolution

So far in this text we have considered general aspects of ecology and particular features of the structure and function of ecosystems, together with the more specific effects of various environmental components. In the preceding chapter we turned again to living organisms and examined the size and growth of the populations that constitute the biotic element of all ecosystems.

The extent to which populations interact with their habitats was first discussed in the second chapter. Such influences, however, extend far beyond the effects on population growth, density, dispersal, and reproduction described there. Habitat factors also determine the nature, that is the *diversity* of living organisms. The extent of genetic variation encountered among the populations and communities of an ecosystem is therefore the result of this environmental interaction. A more usual way of expressing this is by saying that the component species of an ecosystem have *adapted* through the operation of natural selection, under the influence of various interactions within the ecosystem. The diversity of the populations which we encounter in an ecosystem at any given moment of time is the result of these adaptations. The majority of populations in the communities of a contemporary ecosystem are comprised of complex and highly diversified multicellular organisms. Over biological time these have developed directly from the simple unicellular autotrophs which formed the earliest communities three and a half billion years ago.

Charles Darwin, the acknowledged father of evolutionary theory, used the expression "descent with modification" and this phrase very concisely epitomizes this selective genetical aspect of the interaction between the biotic and abiotic elements of ecosystems which results in what we call *evolution*. The essential feature of evolution is that biota vary, and that particular variants receive a disproportionate share of the energy flowing through the ecosystem. This favors their reproduction at a higher rate than other variants, and consequently ensures a greater representation of their particular genotypes in the biotic element. The present chapter is concerned especially with the *origin* of this variation in genotypes and phenotypes, and the manner in which this differential reproduction produces ecological *diversification*.

Variation

Diversity in the biotic element of an ecosystem has many sources. It may result from variation in the genotype arising from *mutation, introgression, recombination,* or *random drift.* The commonest cause of variation is the recombination which arises during sexual reproduction as a result of the random mating of gametes and the random segregation of characters during gamete formation. Entirely new traits can only arise from within a population by gene or chromosome mutation, and from without by introgression of new alleles, following hybridization. Random drift alters the balance of characters but not their nature. Besides these genetical sources, diversity can likewise arise from modification of the phenotype at early or later stages of its development, or throughout its whole period of existence. Even between two individuals of the same population there exists some recognizable variation from one or both of these sources, however slight. Ecologists are forced to classify this observable variation into categories for the pragmatic convenience of handling populations and communities. Without this classificatory need it would be possible to treat the whole of the biotic element as a continuum, and thus avoid the inevitable compromises forced upon us by classificatory requirements.

Phenotypic Versus Genotypic Variation

One form of variation which ecologists can readily investigate experimentally is *phenotypic* variation. This is the diversity of form, function, and behavior that arises as a result of the interplay of limiting substances and factors during the development of individuals with particular genotypes. The earliest method of examining such phenotypic variation was by the *transplant technique.* Plants or animals were removed from their natural habitats and maintained or raised in uniform conditions under a particular set of environmental parameters. In plants, one of the classical experiments using this technique is now referred to as the *California transect,* which is illustrated in Fig. 4-1. Using the transplant technique, Clausen et al. (1948) investigated variation along an E-W transect in central California in populations of species such as yarrow

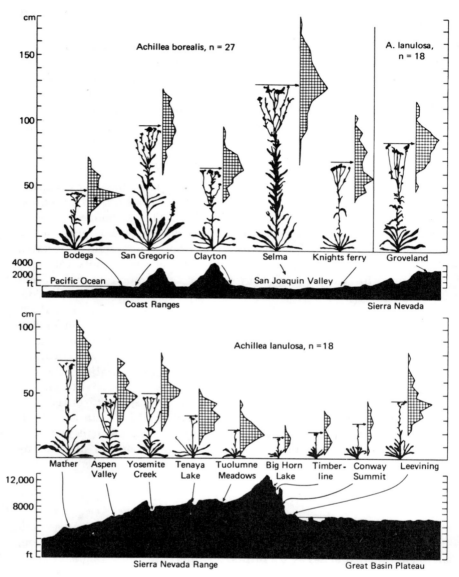

Fig. 4-1. Transplant experiments across a California transect. Ecotypes of several species of *Achillea* grown under uniform conditions in an experimental garden at Stanford University illustrate the degree of genetical variability existing in these various populations. Samples were taken over a 200-mile transect starting on the Pacific coast, traveling east at approximately latitude 38° N to cross the Sierra Nevada, and ending in the Great Basin. The diagram shows the altitudinal variation along the transect, together with the median morphology of the flowering plant and the height variation of each sample of some sixty plants raised in the transplant garden. (After J. Clausen et al., Carnegie Institute of Washington Pub. 81, 1948; reproduced with permission of the publisher.)

(*Achillea*). The transplant technique enabled them to decide how much variation was phenotypic variation, how much variation in the genotype which permitted subdivision of particular populations into series of *ecotypes*.

Another classic study by Mooney and Billings (1961) on alpine sorrel (*Oxyria digyna*) demonstrated the extent to which diversity in physiological as well as morphological characters could be attributable to phenotypic variation, as opposed to genetic variation in the genotypes. This likewise permitted the recognition of various ecotypes.

In animals the phenomenon of phenotypic variation is more commonly described as *acclimation,* and many species have been shown to acclimate in response to various features of the abiotic element of ecosystems. Our own species shows extensive modifications when subjected for example to differences in nutrition in the early years of life, which affects mental development. We also acclimate to different altitudes, which may affect the number of corpuscles in the blood, the capacity of the lungs and the size of the heart. Similar modifications occur when we are exposed to exercise. Some of this phenotypic variation is reversible and represents what is more usually included under the term *acclimation.* Other phenotypic variations such as those which arise through early malnutrition may be irreversible, and should not actually be classified as acclimation. This type of variation has been studied extensively by using identical twins in humans and other species.

Phenotypic variations in animals and plants are not normally taken as a basis for the systematic classification of biological diversity, but only the genetically determined variation, when this can be recognized or presumed. The procedures which are used for this systematic classification are incorporated in *taxonomic principles.* Although certain systematic categories are still regarded as quite controversial, ecologists extensively use these, and so they have to be reviewed here.

Taxonomic and Systematic Categories

The terms taxonomy and systematics are sometimes separately defined, the first to include the methods and principles on which classification is based, the second to embrace the system of classification itself and the material which it covers. Such usage of these two terms is by no means universal, and in America it is presently a more common experience to find the two terms used as synonymous, and covering both these aspects of the classification of biological diversity.

So far in this book the term *species* has been used without further explanation. It is a term ecologists commonly employ without definition, and one moreover which they find indispensable as a reference point in all their ecological work. The particular species definition which is universally accepted today as indicating such a working classificatory category is based on the *typological concept*. This is a little over two centuries old, and has tended to become somewhat stylized and increasingly challenged. Nevertheless the use of this concept has made possible the creation of a system of binomial nomenclature of plants and animals to which all populations can presently be referred, sometimes called *alpha* taxonomy. The morphological or taxonomic species which is created according to this concept is based on one type specimen. This must be identified in some way, its diagnostic characteristics adequately preserved, and its description published, together with its unique name, in a scientific publication. Thus wheat is *Triticum vulgare* L., man, *Homo sapiens* L. The L. refers to the author of the name, in these two cases Carolus Linnaeus (Carl von Linné). The type specimen of *T. vulgare* L. is maintained in the Linnean Herbarium in London, the type of *H. sapiens* L. is said to have been Linnaeus himself. The authority of each specific epithet is usually cited only in taxonomic publications, or where doubt would otherwise exist.

International rules of nomenclature govern separate procedures for plants, animals, and microbes. The primary function of these rules is to ensure stability in the nomenclatural system provided by alpha taxonomy. The use of the rules is not obligatory. Anyone is free to invent his own names, select his own types, and publish his own descriptions. Great difficulty, however, would be encountered in persuading other scientists to take a species name into general usage if it did not comply with these international rules.

This typological system is based on an assumption of the uniqueness and the integrity of the species. In a system first employed by Linnaeus in 1753, each species is incorporated in a *genus* in a binomial system of nomenclature. Genera are grouped further into higher categories, of which the most commonly used are successively *families, orders*, and *classes*. Approximately 300,000 species of plants have been described and nearly one million species of animals. This does not include all living populations, perhaps 10 to 20 percent more plant species should be added and rather higher percentages of animals. Some workers consider that only about half the existing species of insects for example have yet been described. Nor are fossil species represented in this typological system in anything more than a fragmentary manner.

Biological Species

Increasing emphasis on genetical studies which began about 1940 led to the development of the concept of what is generally known as the *biological species*. This supposes that taxonomic variation is not continuous, but that there are discontinuities corresponding with discrete breeding units. Members of such units freely exchange genetical material with one another, but do not readily do so with individuals in other units. The discrete breeding entities represent *biological species*.

There are in actuality many such biological species, and these do indeed have breeding discontinuities which prevent their readily exchanging genetic materials with other biological species. The more examples are investigated however, the more apparent it becomes that the biological species is unacceptable as a *universal* taxonomic unit, and such genetically determined taxonomic entities may be proved eventually to be a small minority. In any case there is no assemblage of biological species in any way approaching the faunal and floristic listings of taxonomic species established using the typological concept.

Infraspecific Categories

Ecologists frequently encounter populations having somewhat more restricted variation than that presented by the whole range which a taxonomic or biological species is known to exhibit. Any such taxonomic entity whose nature is not to be specified may be referred to as a *taxon* (plural *taxa*). For animals the term in most frequent current use for such an infraspecific category is *deme*. This is a local animal population whose members freely interbreed among themselves, but not with members of other demes. For some purposes the older category of *subspecies* is still retained (Fig. 4-2). Usually a subspecies is a geographical variant of a taxonomic species, as for example *Bison bison bison* for the plains bison, *Bison bison woodsii* for the wood bison (Fig. 4-3). Plant species are also commonly divided into geographic subspecies.

Brief mention has already been made of the plant taxonomic category which ecologists most commonly utilize, the *ecotype* (Fig. 4-4). This is a population which is encountered in particular and definable habitat conditions and exhibits a limited portion of the total variation of a species. Most species which occur on both serpentine and nonserpentine soils for example have recognizable *ecotypes* which occur on the one or the other soil type. Plant species which occur across gradients of limiting factors, like those having a dispersal area with increasing latitude and corresponding photoperiod changes, can commonly be fragmented into ecotypes with different photoperiodic responses.

Of frequent occurrence in plants, but very much less in animals, is the category known as the *cytotype*. This is a variant recognizable by its different *karyotype* or chromosome compliment. The haploid number may for example be doubled as in tetraploids, trebled as in triploids, six times the basic haploid number as in hexaploids, or vary by the duplication or loss of one or several chromosomes as in aneuploids. Associated

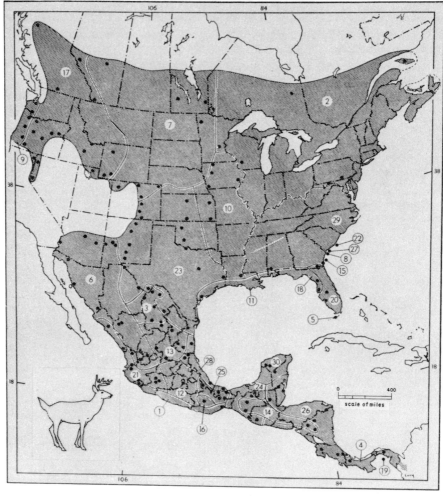

Odocoileus virginiana

Guide to subspecies	7. O. v. dacotensis	15. O. v. nigribarbis	23. O. v. texans
	8. O. v. hiltonensis	16. O. v. oaxacensis	24. O. v. thomas
1. O. v. acapulcensis	9. O. v. leucura	17. O. v. ochroura	25. O. v. toltecs
2. O. v. borealis	10. O. v. macroura	18. O. v. osceola	26. O. v. truei
3. O. v. carminis	11. O. v. mcilhennyi	19. O. v. rothschildi	27. O. v. venateria
4. O. v. chiriquensis	12. O. v. mexicana	20. O. v. seminola	28. O. v. veraececis
5. O. v. clavia	13. O. v. miquihuanensis	21. O. v. sinaloae	29. virginiana
6. O. v. couesi	14. O. v. nelsoni	22. O. v. taurinsulae	30. O. v. yucatanensis

Fig. 4-2. Subspeciation in American mammals. A map showing the range of localities of thirty of the subspecies which have been described for the white-tailed deer (*Odocoileus virginianus*). Somewhat similar numbers of subspecies have been erected for other widely dispersed mammals such as the white-footed deer mouse (*Peromyscus maniculatus*), and for various birds and other animal groups. Such infraspecific categorization has little ecological significance other than to indicate that a given animal is a sample of a deme of a particular species established in a given locality at a given moment of time. (After Hall and Kelson, *The Mammals of North America,* © 1959, reproduced with permission of the Ronald Press Company, New York.)

Fig. 4-3. A southern race of Burchell's zebra (Equus burchelliae) showing the characteristic striping of this geographic form. Other geographic races of zebra occur in other parts of Africa, each with its distinct patterning. These geographic races are usually allopatric, but in a few instances their dispersal areas overlap. In such areas, hybrids are believed not to occur, and interbreeding has not been observed. It is thought that the different patterning characterizing each race fails to stimulate release mechanisms necessary to evoke mating behavior in the stallions when these are of a different race from the mares.

with each cytotype there may or may not be some readily observable physiological and morphological variation.

Interest in karyotypes, together with an increasing emphasis on *experimental* methods in taxonomy, led to *biosystematic* studies and the *biological species,* but the challenge to traditional taxonomy did not stop there. Many workers now insist that the ultimate or omega-taxonomy will have regard for *all* the characters which organisms of a population express, and that none should have any special emphasis. These ideas established the approach which has come to be known as *numerical taxonomy.*

Numerical Taxonomy

While the principles of numerical taxonomy were first expounded in 1963 by Sokal and Sneath (1966) they are based on ideas presented by the eighteenth-century French botanist Michel Adanson. He considered that similarity between any two taxa was a function of the totality of

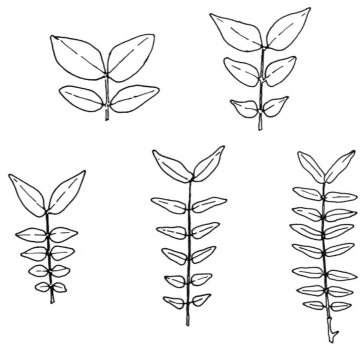

Fig. 4-4. Ecotypic variation in plants is sometimes expressed in the form of an obvious morphological differentiation. In the central African leguminous tree species *Brachystegia spiciformis* illustrated here, the number of leaflets per leaf can be shown to be positively correlated with the mean annual rainfall of the sampling area. This continuum of leaflet increase along the rainfall gradient must be associated with other more cryptic and often physiological adaptations. Insufficient sampling led initially to the establishment of a number of distinct species representing particular sections of this continuum.

characters which could be compared, and that this affinity was independent of phylogenetic considerations. Sokal and Sneath maintained that taxonomy was an empirical science, that similarity matrices should be prepared from an examination of a minimum of 60 characters, and that hierarchical dendrograms showing degrees of similarity could be constructed to illustrate phenetic relationships between taxa.

The necessary computational exercises for such operations required the use of electronic data processing with a digital computer, which of necessity would limit the general introduction of numerical taxonomic procedures even if they were to be completely accepted. Nevertheless the ideas on which numerical taxonomy is based can be said to have found general, if sometimes conditional agreement.

Whatever the nature of the classificatory units employed for categorizing populations of plants, animals, and microbes, they are all basically determined by and attributable to differences in gene frequency between individual organisms.

Gene Frequencies

Within a given deme or ecotype population the sum total of alleles at each individual chromosome locus, when computed as a percentage figure for the total population, provides an expression of the *gene frequency* of each particular allele. Whatever its frequency, each allele has arisen in the past by *mutation*. Unless further mutations or a back mutation are occurring, or selection is favoring or lowering the representation of particular alleles in successive generations, this gene frequency for the particular allele will remain unchanged. This circumstance has been called the *Hardy-Weinberg law*. If there is any change in gene frequencies as a result of selection pressures or because of the occurrence of mutation, then there is a disturbance in the Hardy-Weinberg equilibrium. The maintenance of the same frequency for an allele over a period of time in a given population is therefore assumed to be an indication of the absence of mutation and selection pressure. Contrariwise, a change of gene frequency is assumed to indicate the occurrence of mutation, or selection pressure or both, except when random drift occurs.

Selection Pressures

The vast majority of mutations are initially expressed as recessive characters. They are therefore not at first subjected to selection pressure, because they are not discernible phenotypic traits until their gene frequency reaches a sufficiently high value in the population for heterozygous parents to have any real probability of mating, and thus producing homozygous recessive offspring revealing the trait. When mutations are finally expressed in this homozygous form, they add to the range of genetic variation which has arisen in a population as a result of recombination, introgression, and random drift.

The selection pressures operating on the variants in a population arising from such sources will do so with differing intensities and through several modes of operation. If for example the genetic variation produces phenotypes whose range of tolerance for particular factors lies outside the particular parameters of the habitat, such individuals will probably die before they reproduce. This is the form of selection which is most likely to occur in individuals of populations of invertebrate or plant groups, where there are very heavy mortalities among embryonic and juvenile forms. It is less likely to operate in populations of such groups as mammals, which as shown in Figs. 3-5 and 3-6 have a different form of survivorship curve.

Alternatively among individual variants surviving to maturity, selection may favor increased reproduction of particular individuals, or result in a reduced reproductive rate. In this instance, selection pressures are operating through a differential reproductive factor. This appears to be a very common mode of operation among species that do not have excessive juvenile mortalities. Darwin described the individuals with the higher reproductive rates are being more "fit" than those individuals with a lower one, which were less "fit."

In either case, the result of the operation of natural selection on a population is to maximize "fitness." Such selection for fitness has two consequences, which may be independent or linked. It may create new patterns of variation in a population, and it may reduce the diversity of genotypes encountered within it. In domesticated animals and plants, and frequently in laboratory experimentation, artificial selection has often been employed to achieve both these objectives.

The Experimental Demonstration of Natural Selection

Many experiments with fruit flies (*Drosophila*), bread molds (*Neurospora*) and other such species commonly used in genetics laboratories, illustrate the operation of natural selection. In one such experiment two populations of *Drosophila mulleri* were caged together. One had the normal genotype for this species, the other being homozygous for a mutant allele previously induced by exposure to X-ray radiation, producing a chromosome inversion. The two populations introduced together into the cage mingled to form a single population which had 15 percent individuals homozygous for the normal genotype, 85 percent homozygous for the mutant. Food was continuously replenished in the culture, and the population followed a sigmoid form of growth and became stabilized at the asymptote. At regular intervals egg samples were withdrawn from the culture and allowed to form larvae. The gene frequencies for the normal and for the mutant form of chromosome were then determined by examining the salivary-gland chromosomes of these larvae. The gene frequencies of the mutant form obtained from these samples is plotted in Fig. 4-5. As can be seen from this graph, in less than six months—after about 20 generations of fruit flies—this frequency had fallen from 0.85 to 0.0075.

In this experiment, despite the fact that all other known influences beyond the single mutant one had been eliminated, the Hardy-Weinberg

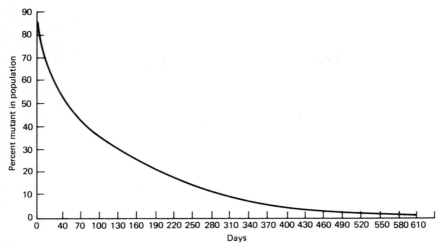

Fig. 4-5. Changes in gene frequency in a laboratory population of fruit fly. In this experiment a closed population of *Drosophila mulleri* initially with an 85 percent mutant frequency experienced a decline to 0.75 percent frequency in this mutant in 610 days after some 20 generations. Presumably the individuals possessing the mutant allele are not as viable as the "normal" type, which is favored by natural selection.
See text for a further explanation. (Data extracted from Mettler and Gregg, 1969.)

equilibrium between the two genotypes was not maintained. Some selection pressure therefore had to be operating which caused the mutant genotype to be represented in successive generations with lowered gene frequencies. In Darwinian terms, the lower reproductive capacity of the mutant form made it less "fit" under these particular experimental conditions.

Types of Selection

The operation of selection pressures within a single population as described in this experiment has been separated into three types—*stabilizing, directional,* and *disruptive.* These three types are illustrated diagrammatically in Fig. 4-6.

Stabilizing selection occurs when abiotic (habitat) features and factors remain constant. It favors those genotypes that develop phenotypes optimally suited to this particular fixed set of conditions. It will tend to eliminate or produce lower reproductive rates in phenotypes which are not so optimally suited, and thus reduce the representation of their particular genotypes in the population. An example of stabilizing selection is evidenced in the birth weight of human babies (Karn and Penrose, 1966). The optimum birth weight at least in industrialized societies appears to lie at approximately 8 lb at birth. The greater the variation

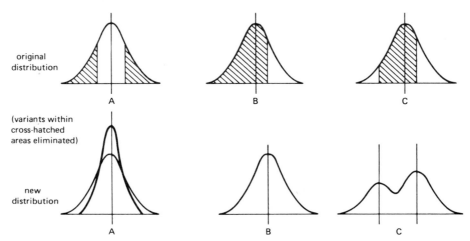

Fig. 4-6. The three basic modes of selection and their effects on genetic variation within a population. A is known as *stabilizing,* B as *directional,* C as *disruptive* selection. Graphs represent the extent of variation in a given character along the horizontal axis, and the number of individuals with particular values within this variation on the vertical axis. The vertical lines represent the mean value in each instance.

In A, selection for variants lying close to the mean prevents spread of the curve. In B, selection of variants lying to the extreme right of the mean causes a directional shift in the mean. In C, elimination of variants lying close to the mean produces a bimodal distribution curve.

away from this weight in either direction the more reduced the survival rate in the first month of life.

Directional selection results when the mode of the phenotypic optimum does not coincide with that of the parental one. There are many examples of this. That which occurs in relatively small populations and is given the designation of *random drift* or *genetic drift,* is most frequently described. Glass (1954) investigated variations in the various biochemical features of the blood of the Dunkers, a sect of German immigrants that has since its establishment in several parts of eastern America largely interbred within its own population. In three generations examined, the gene frequencies for the MN blood series were 0.55M and 0.45N in the oldest (identical with the surrounding non-Dunker populations) 0.66M and 0.34N in the middle generation, and 0.735M and 0.205N in the youngest. Directional selection of the kind described as genetic drift was therefore quite rapidly altering the gene frequencies of the MN blood series in this human population.

Disruptive selection can be regarded as the contrary of stabilizing selection in that it tends to *diversify* the range of genotypic variation in a population. This may result in the appearance of two phenotypic optimum modes, phenotypes lying between the modes as well as beyond them being selected against. Experimental simulation of disruptive selection has been demonstrated by Thoday and Gibson (1962) in laboratory experiments on bristle numbers in fruit flies. They were able artificially to select simultaneously for both high and low bristle numbers within the same population.

Disruptive selection is not restricted to bimodal situations, but within the same population a number of different modes may exist simultaneously, a condition described as *polymorphism*. A well-known example of this is provided by the African swallow-tailed butterfly (*Papilio dardanus*) (Fig. 4-7). This nonpoisonous species mimics several forms of

Fig. 4-7. Polymorphism resulting from disruptive selection. Monarch butterflies (*Danaus plexippus*) on left are unpalatable to bird predators. Viceroy butterflies (*Limenites archippus*) on right which mimic them are edible, but generally escape predation because of their mimicry of the Monarchs. Viceroys are polymorphic because such adaptation minimizes the ratio of palatable to unpalatable forms.

distasteful butterflies thereby reducing losses from bird predation. Selection for such mimicry will not be high if the number of mimics is comparable with the population size of the distasteful species which is being mimicked. If on the other hand the same population is polymorphic for mimicry—that is, if it contains a number of different genotypes, each producing phenotypes mimicking one or another different species of distasteful butterflies—the size of the mimic population relative to the distasteful population is considerably reduced, and selection pressures favoring the mimic will be correspondingly higher.

Selection pressures are again considered in the next chapter in regard to the niche concept, niche differentiation, and character displacement. They are also discussed when mimicry is examined in relation to secondary plant substances (*primary* plant substances are those universal metabolites such as proteins, sugars, fats, vitamins).

Selection under Natural Conditions

The classical demonstration of the operation of natural selection in wild populations was performed using melanistic forms of moths in industrial and nonindustrial districts of England by Kettlewell (Table 4-1). Stated very briefly in simplified form, dark forms of some of these previously light-colored moths began to be noted toward the end of the last century in industrialized areas of Britain, and their numbers were observed to be increasing. Simultaneously with this, lichens which are extremely sensitive to air pollution by industrial contaminants dis-

TABLE 4-1

Natural selection of melanistic moths—a classical demonstration of selection operating under natural conditions. *Industrial melanism* in moths developed during this century in Britain as a result of habitat changes induced by air pollution. Trees in industrial areas previously with light-colored bark provided instead the darker surface of the exposed bark. Air pollution had eliminated the previous lichen cover. Pigmented forms of moths resting on the dark tree trunks of industrial areas were less heavily predated by birds than the original light colored moth forms, as shown in this experiment reported by H. B. D. Kettlewell.

	Industrial Area (near Birmingham) Dark tree-trunks		Rural Area (Dorset) Light tree-trunks	
	Original Light Forms	Dark Mutants	Original Light Forms	Dark Mutants
Total number of moths released	154	64	473	496
Number of moths not recaptured ..	72	48	443	434
Number of moths recaptured	82	16	30	62

Data extracted from Kettlewell, *Annual Review of Entomology* 6:245–262, 1961.

appeared from the bark of the trees in industrial areas. Heavy growths of lichen had previously covered tree trunks extensively, and generally provided a light background on which moths alighted. The lichen-free bark surfaces were now generally darker. Kettlewell was able to demonstrate by capturing both light and dark forms of moths, marking them, releasing them, and subsequently counting the numbers recaptured that heavier losses occurred among the light forms in industrial areas, and among the dark forms in nonindustrial ones (Kettlewell, 1961). He was also able to confirm that differential loss was due to a differential predation. The more conspicuous nature of the dark moths in nonindustrial areas and of the light moths in industrial ones caused this differential predation.

Ecological Significance of Selection Pressure

The result of the operation of selection acting in these several ways is to produce what is known as *adaptation*. This is the creation of populations whose gene frequencies are continuously modified so that their phenotypic expression produces a population optimally adjusted to the prevailing environmental features of particular ecosystems. Where these abiotic features are undergoing rapid change, new forms are evolving by adaptation which often have to be classified as new taxa. In other words, *speciation* is occurring. Both natural and artificial selection have played and will continue to play a major role in the development of new taxa which have economic significance for human societies. Conspicuous among these are pesticide-resistant pests, and mutant forms of diseases and pests which can attack previously resistant economic crops.

Speciation

Traditionally it has been considered that selection pressures and subsequent changes in the gene frequencies of particular populations have been insufficient *by themselves* to bring about speciation. It has been postulated that some measure of *isolation* must also exist. The most obvious form of isolation is the *geographical* separation which occurs between the allopatric segments of a once sympatric population. The existence of infraspecific populations at the subspecies level is commonly attributed to some measure of geographical separation, as illustrated by Figs. 4-2 and 4-3.

Geographical separation cannot, however, account for the *sympatric* fragmentation of a species population. Various mechanisms of a morphological, phenetic, physiological, cytological or ethological nature have accordingly been postulated to explain such apparent systematic discontinuities (Table 4-2). Recently however there has been a tendency to

TABLE 4-2

Isolating mechanisms tending to separate sympatric species and infraspecific populations. While allopatry by itself is sufficient to provide some degree at least of genetic isolation between populations, such geographic isolation obviously cannot operate in the case of sympatric populations. Population geneticists have detected a number of mechanisms which may contribute to the separation of sympatric species and subspecific groups. These may be classified into the following categories. (See text for further comment.)

A. Barriers to Interbreeding
    i.   Separation of breeding or flowering seasons
   ii.   Behavioral separation of breeding individuals
  iii.   Structural adaptations preventing interbreeding
   iv.   Physiological incompatibilities preventing fertilization or successful embryonic development
   v.   Cytological incompatibilities preventing fertilization or successful embryonic development

B. Barriers to survival of hybrids
    i.   Expression of lethal or disadvantageous traits
   ii.   Sterility in hybrid individuals
  iii.   Either (i) or (ii) or both in second and subsequent generation hybrids

minimize the effects of such isolating mechanisms (Fig. 4-8). It is suggested instead that selection pressures themselves, rather than any specific isolating mechanism, maintain apparent differences between related sympatric populations (Ehrlich and Raven, 1969).

Such considerations lead us back into the same taxonomic-biological species argument already examined earlier in this chapter. From a pragmatic ecological viewpoint, breeding barriers between both allopatric and sympatric populations may be identified in many instances (Table 4-2). The populations thus isolated to a greater or lesser degree can be treated as species in the traditional understanding of this term. It can be said that speciation has occurred, an ecological explanation for this differentiation can be presented, and the ecological reactions and functions of the species thus established can be described.

Adaptive Radiation and Convergent Evolution

The total adaptation which occurs in the component populations of particular communities in given ecosystems is not usually proceeding in a completely randomized and totally disconnected series of directions. More commonly evolution is moving, as Darwin originally envisaged, by small stages in particular directions. When these are tending to *reduce* the diversity exhibited by the populations within a given community, we speak of *convergent evolution,* when these are leading to a *greater* diversity, we speak of *adaptive radiation.*

Fig. 4-8. Selective effects of the microenvironment may develop particular variants of plants and animals, rather than causing their complete disappearance from a given area. On sea coasts, where salt spray formed by breaking waves is blown inland, plants in the coastal ecosystems are liable to suffer severe damage from the salt deposited on their surfaces. Prostrate variants of plants, which largely escape exposure to such salt concentrations because of their low profile, therefore tend to be selected as against upright forms. This photograph shows such a microecosystem of prostrate variants on the coast of Ghana in West Africa.

Convergent evolution leads to the closer approximation of producer, primary consumer, secondary consumer, or reducer dominants in particular ecosystems in their morphological, physiological, and behavioral characteristics. We have already noted that the contemporary dominant herbivores of the Californian chaparral, are (ignoring deer) seed-eating nocturnal rodents, with cheek pouches, living in burrows, surviving if necessary on metabolic water alone. The commoner producer dominants, with such varied systematic origins as Asteraceae, Polygonaceae, Anacardiaceae and Rhamnaceae, are deep-rooted, medium-sized evergreen shrubs with small, leathery, nonentire leaves, and often primarily bird-distributed propagules. Convergent evolution may have brought these dominants to such an extent of morphological similarity that they carry the same specific epithet, as in the case of *Eriogonum fasciculatum* (Polygonaceae) and *Adenostoma fasciculatum* (Rosaceae) (Fig. 4-9). The most familiar example among flowering plants is the number of families that have *succulent* forms adapted to desert conditions.

Adaptive radiation occurs especially when some major ecological

(a)

(b)

Fig. 4-9. Convergent evolution in plants can produce species morphologically so similar that they are given the same or a closely similar specific epithet. In chaparral A, "chamise" (*Adenostoma fasciculatum*—Rosaceae) and B, "buckwheat" (*Eriogonum fasciculatum*—Polygonaceae), although classified in widely separated families, very closely resemble one another morphologically. This convergent evolution has even resulted in the adaptation of the leaves into bundles. Hence the particular specific epithet adopted, which derives in each case from the "fasces" or bundle of faggots carried as an emblem of office by Roman magistrates.

change occurs. In geological time this happens when vast areas of virgin habitat appear—for example when the morainic material deposited by glaciers is exposed as they retreat, or when lowering of sea levels to uncover a continental shelf provides a land bridge from one continent to another. On a smaller scale, adaptive radiation often proceeds amongst the earliest invaders of new volcanic areas, and especially islands. The traditional example of the first type is provided by Australian marsupials. After the arrival of these early mammalian forms in Australia, that continent became isolated by land approach until the Pleistocene Ice Ages, consequently no *eutherian* (placental) mammalian forms became established there. Marsupial forms accordingly underwent *adaptive radiation,* evolving by niche diversification *ecological equivalents* very similar in forms and functions to the more advanced mammalian dominants of comparable ecosystems on other continents.

Another very frequently quoted example of adaptive radiation, which is of the second type is "Darwin's finches" referring to the speciation encountered in this bird group on the various islands of the Galapagos archepelago (Fig. 4-10). On the South American mainland, finches constitute a single group of closely related insectivorous birds. Supposedly following invasion of the new habitats which the volcanic Galapagos islands once represented, the finches evolved three ground-dwelling species, either seed-eating, or living on cactus, an arboreal vegetarian species with a parrotlike beak, a "woodpecker" type using a cactus spine to skewer out insect larvae, and a number of "warbler" species.

Adaptive radiation produced other communities very familiar to us, like the game antelopes of the African savannas and the various monkey species of the Old and New World tropical rain forests. Artificial selection at first empirical, now along planned breeding programs, enabled us to simulate adaptive radiation to evolve domestic forms of animals and plants best adapted to the synthetic or degraded ecosystems whose productivity we wished to exploit for the support of our own species. Some of the problems encountered in such applied ecological procedures are briefly reviewed by Pirie (1969).

In this chapter we have reviewed the systematic categories with which ecologists have to work, and the habitat factors which influence them and lead to further population adaptation and evolution. Such considerations as we have reviewed here are not however the sum total of evolution. We have examined only the effects of abiotic factors in evolution. Biotic interactions may and usually do play an equally important role in this process. The significance of these and other interactions between populations is now considered in the next chapter.

Islands	Types of finch					
CENTRAL GROUP	1	4	3	5		2
Tower	1			4	3	
Hood		4			1/3	
Wenmann	1			3/4		
Culpepper				3/4	1	

(a)

Key

1   large ground finch (magnirostris)

2   medium ground finch   (fortis)

3   cactus ground finch (conirostris)

4   small ground finch (fuliginosa)

5   humid ground finch (difficilis)

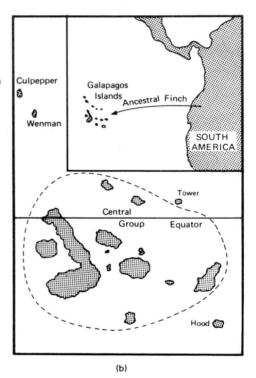

(b)

Fig. 4-10. The Galapagos finches first described by Charles Darwin and more recently studied by David Lack, provided a classic example of speciation by adaptive radiation in the absence of competition. The island group are not strong flyers, so that distribution of island species is quite local (b). The simplified diagram shows here how in the absence of competition the ecological niche of a given finch species may be wider on one island than on another (a). Where competition is absent one taxonomic species may function in several different ecological roles. Thus *Geospira difficilis*, a large ground finch in the central group of islands, plays the role of both a cactus ground finch (3) and a small ground finch (4) in Culpepper, in the absence of *G. conirostris* and *G. fuliginosa*. The nature of competition between species as illustrated both here and in Fig. 4-9 is described more specifically in Chapter 5.

Bibliography

References

Camin, J. H., and P. R. Ehrlich, "Natural Selection in Water Snakes (*Natrix sipedon* L.) on Islands in Lake Erie," *Evolution* 12:504–511, 1958.

Carson, H. L., "Genetic Conditions Which Promote or Retard the Formation of Species," *Cold Spring Harbor Symposia on Quant. Biol.* 24:87–105, 1959.

Clausen, J., D. D. Keck, and W. M. Hiesey, "Experimental Studies on the Nature of Species: III. Environmental Responses of Climatic Races of *Achillea*," *Carnegie Institute of Washington Publication No.* 581, 1948.

Cody, M. L., "A General Theory of Clutch Size," *Evolution*, 20:174–184, 1965.

Deevey, E. S., "Specific Diversity in Fossil Assemblages," in *Diversity and Stability in Ecological Systems*, Brookhaven Symposia in Biology No. 22:224–241, 1969.

Ehrlich, P. R., and R. W. Holm, "Patterns and Populations" *Science* 137:652–657, 1962.

Ehrlich, P. R., and P. H. Raven, "Differentiation of Populations," *Science* 165:1228–1232, 1969.

Hamilton, T. H., *Process and Pattern in Evolution*. New York: Macmillan, 1967.

Johnston, R. F., and R. K. Selander, "House Sparrows: Rapid Evolution of Races in North America." *Science* 144:548–550, 1964.

Kettlewell, H. B. D., "The Phenomenon of Industrial Melanism in Lepidoptera," *Annual Review of Entomology* 6:245–262, 1961.

King, J. L., "Continuously Distributed Factors Affecting Fitness," *Genetics* 55:483–492, 1967.

Lack, D., "Evolutionary Ecology," *Journal of Ecology*, 34:223–231, 1965.

———, "Darwin's Finches," *Scientific American*, 188(4):66–72, 1953.

Love, A., "The Biological Species Concept and its Evolutionary Structure," *Taxon*, 13:33–44, 1964.

Mayr, E., "Isolation as an Evolutionary Factor," *Proc. Amer. Philosophical Soc.*, 103:221–230, 1959.

Macmillan, C., "Ecotypes and Community Function," *American Naturalist* 94:246–255, 1960.

Milkman, R. D., "Heterosis as a Major Cause of Heterozygosity in Nature," *Genetics* 55:493–495, 1967.

Mooney, H. A., and W. D. Billings, "Comparative Physiological Ecology of Arctic and Alpine Populations of *Oxyria digyna*," *Ecological Monographs* 31:1–29, 1961.

Sanders, H. L., "Marine Benthic Diversity: a Comparative Study," *American Naturalist*, 102:243–282, 1968.

Simpson, G. G., "The First Three Billion Years of Community Evolution," in *Diversity and Stability in Ecological Systems*, Brookhaven Symposia in Biology No. 22:162–177, 1969.

Sokal, R. R., and P. H. A. Sneath, "Efficiency in Taxonomy," *Taxon* 15:1–21, 1966.

Stebbins, G. L., "The Experimental Approach to Problems of Evolution," *Folia biologica* (Prague) 11:1–10, 1965.

———, *Processes of Organic Evolution.* Englewood Cliffs, N.J.: Prentice-Hall, 1966.

Thoday, J. M., and J. B. Gibson, "Isolation by Disruptive Selection," *Nature*, 193:1164–1166, 1962.

White, M. J. D., "Models of Speciation," *Science* 159:1065–1070, 1968.

Further Readings in Human Ecology

Allen, G. E., "Science and Society in the Eugenic Thought of H. J. Muller," *Bioscience* 20:346–353, 1970.

Brown, A. W. A., "Insecticide Resistance Comes of Age," *Bulletin of the Entomological Society of America* 14:3–9, 1968.

Caspari, E., "Selective Forces in the Evolution of Man," *American Naturalist* 97:5–14, 1963.

Chandler, R. F., *New Horizons for an Ancient Crop*, XIth Intern. Bot. Congr., Seattle, Wash., All-Congress Symposium, World Food Supply, Aug. 28, 1969.

Crow, J. F., "Some Possibilities for Measuring Selection Intensities in Man," *Human Biology* 56:30:1–13, 1958.

Dobzhansky, T., "The Present Evolution of Man," *Scientific American* 203(3): 206–217, 1960.

Glass, H. Bentley, "Genetic Changes in Human Populations, Especially Those Due to Gene Flow and Genetic Drift," *Advances in Genetics* 6:95–139, 1954.

Mourant, A. E., "Human Blood Groups and Natural Selection," *Cold Spring Harbor Symp. Quant. Biol.*, 24:57–63, 1959.

Review Questions

1. What is the difference between *genotypic* and *phenotypic* variation within populations? Describe the commoner causes of phenotypic variation and their ecological significance.

2. Describe the causes of genotypic variations in animal and plant populations. Illustrate your answer by reference to named examples.

3. State what is meant by the Hardy-Weinberg law. Discuss the ecological significance of this law.

4. Describe what you understand by *speciation*. Discuss the extent to which this is dependent on genotypic variation.

5. How is genotypic variation maintained in a species population?

6. Discuss the several concepts of the nature of species. State which one of these you would consider most applicable in ecological work and provide your reasons for selecting this particular one.

7. What types of selection pressure are usually recognized? Illustrate each of the types you mention by reference to specific examples.

8. Describe the difference between *adaptive radiation* and *convergent evolution*. Discuss the significance of these two processes in macroevolution.

9. Recent views minimize the significance of breeding barriers in speciation. Outline the argument for supposing that selection pressures have played a more significant role in the differentiation of species than was formerly considered possible.

5

Population Interactions

In the preceding chapter we considered the effects of *abiotic* factors on population evolution. In order to complete the survey of environmental interactions it is necessary to consider also the *biotic* influences on population evolution exerted by other members of the same communities. In our previous discussions of ecosystems it was noted that they included two interfaces at which populations interacted vigorously. The one was between populations at the same trophic level, which leads to *competition*. The other was between one trophic level and the immediately succeeding one which utilizes the first as a food resource. This second interface displayed such relationships as *predation* and other forms of consumer-consumed interdependence. The interactions at both these types of interface modify the nature of evolutionary processes as well as influencing such features as population dynamics, energy flow and utilization, and the diversity and stability of ecosystems. The survey of the interaction between populations and their environment is therefore completed in this chapter by reviewing these several features of biotic interaction, and especially the nature of *competition.*

Competition

Competition is a term that is variously interpreted in ecology. In its broadest sense it may be defined as the extent of interaction which develops between two sympatric populations—that is, between two populations occurring in the same ecosystem. As already noted, when these populations occur at the same trophic level there are two possibilities. Either one of these populations will pass to extinction, or else one or both will be so modified by selective adaptation as to minimize the extent of the competition between them.

Ecological concepts of the nature of this type of competition are derived from classical theoretical considerations developed independently by Lotka and Volterra, and from experimental demonstrations of the validity of these theoretical models by Gause. Hence the law governing competition has been extensively known as the "Rule of Gause" (Fig. 5-1). A more recent statement of such principles has been incorporated in the *competitive exclusion principle* (Hardin, 1961).

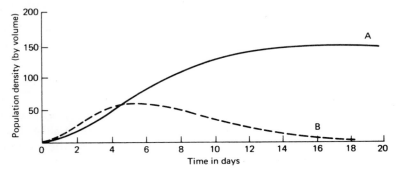

Fig. 5-1. The rule of Gause competition between two species populations of Paramecium in a closed culture. *P. caudatum* (B) declines to extinction as a result of its inability to compete with *P. aurelia* (A) for the same food resources. (Redrawn: based on Gause, 1934.)

The Competitive Exclusion Principle

The idea of competitive exclusion may be developed from the standard differential equation for exponential population growth given in Chapter 3 viz.:

$$rN = \frac{dN}{dt}$$

It was noted there that some environmental factor usually intervened eventually to reduce r and prevent exponential growth continuing. This factor is commonly expressed as K, the carrying capacity, as was seen. This carrying capacity factor K, whether it is a shortage of food, of space, or any other limiting factor, finally reduces r to zero.

If two species are utilizing the same resource, one will most probably reach this position where its growth rate is zero before the other. Any further continuation of population growth in the second species can then only be achieved by reducing the r of this first species to a negative value. The second species is then said to have a "competitive advantage" over the first.

If this competitive advantage of the "fitter" species continues, the negative growth rate—that is, the progressive failure of the first species to maintain its population size—finally causes it to pass to extinction.

In actuality before this happens the situation is likely to change. When the second species is so reduced in population size it becomes rare, its growth rate r seems to pick up again, for several different reasons. Principally perhaps this is because individuals of the rare population are so widely dispersed that they cease to compete with one another for the limited food resource.

Experimental Demonstration of Competition

Experimental confirmation of the findings of Lotka, Volterra, and Gause came with a number of now classic experiments which demonstrated under laboratory conditions the reality of the competitive exclusion principle. The best known of these involved a long carefully planned series of experiments by O. and T. Park some two decades ago using related species of flour beetles, especially *Tribolium confusum* and *T. castaneum*. The results of two fairly typical experiments in the series are illustrated in Fig. 5-2.

This work also stressed the separation which can be made between two components of competition, *exploitation* and *interference*. *Exploitation* occurs when two or more populations draw upon the same limiting resource. *Interference* is competition defined in a more restricted sense, as the interaction between two or more populations which disturbs their growth or influences their survival. As will be discussed later, chemical interactions of an antibiotic or allelopathic nature occur between competing species, and result in one population reducing the rate of growth of another. This is *interference,* which is quite different from one species competing with another for the same limited resource such as food or space. The nature of *exploitation* and *interference* are further illustrated in a model prepared by T. Park (1962) shown in Fig. 5-3.

Harper (1961) and his co-workers initiated experiments which similarly demonstrated the existence of competition under experimental conditions in plants. In one such experiment (Fig. 5-4) they used three species of duckweed (*Lemna*) and one of water fern (*Salvinia natans*) which were cultured both separately and together. When *L. polyrrhiza* was grown with either *S. natans* or *L. gibba* the population growth of these two species was drastically reduced. It seemed likely if the experiment had been carried for a sufficient period they would eventually have been eliminated.

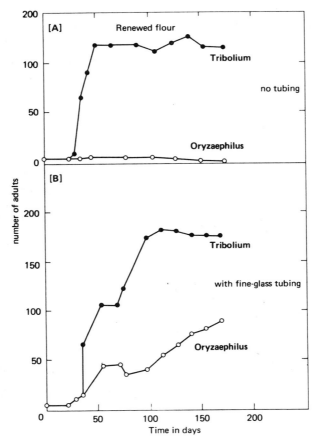

Fig. 5-2. Competition between two species of flour beetle. In A, *Tribolium* eliminates *Oryzaephilus* from a closed culture in 87 percent of the replicates. In B, a beetle population of the genus *Tribolium* coexists with another beetle population of the genus *Oryzaephilus* when fine glass tubing is added to the culture. The opportunity for niche-differentiation which is offered with the addition of the tubing lessens the intensity of the direct competition between the two populations. (After Park, 1954, and Crombie 1946; reproduced with permission of the publishers.)

Competition under Natural Conditions

The investigation of competition between populations in natural ecosystems presents even more difficulty, because of the complexity of the situation in which there are a number of populations in the natural community and a variety of abiotic substances and factors in the un-

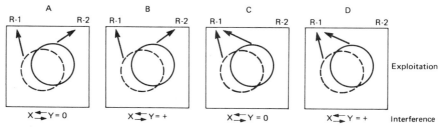

Fig. 5-3. An abstract model of competition illustrating the *interference* and *exploitation* elements of competition between two species populations X and Y. In case A species X utilizes resource R–1, species Y, resouce R–2; there is no *interference,* competition is absent. In case B there is interference between the species, which still utilize different resources. In case C there is no such interference, but there is competition because both species now *exploit* the same food resource. In case D there is both *exploitation* and *interference;* competition is intense. (Redrawn based on Park, 1962.)

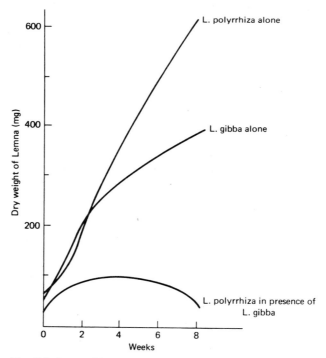

Fig. 5-4. Competition between two species of duckweed—the growth of *Lemna polyrrhiza* and *L. gibba* separately and together. Grown in mixed culture *L. polyrrhiza* suffered competitive exclusion. (Data extracted from Harper, 1961.)

controlled habitat. An early and successful attempt to demonstrate the existence of competition between species under natural conditions has been described by Connell (1961) who studied the interaction between populations of two barnacle genera *Balanus* and *Chthamalus*. Species belonging to these two genera occur in typically restricted zones of temperate rocky coasts, and Connell demonstrated that their zonational relationships resulted from interspecific competition.

A number of special features of barnacles made them very suitable for such an experiment. Once they become attached as juvenile forms, barnacles spend the rest of their lives anchored in the same position. The cover and effective position of *individual* barnacles may thus be followed. In Scotland, where this work was done, the zone of the *Balanus* species was lower than that of the *Chthalamus* species. This did not, however, prevent young forms of *Chthalamus* colonizing the lower zone, although when they became established there they did not grow as rapidly as young forms of *Balanus*. The contrary was not true: young forms of *Balanus* did not colonize the upper zone occupied by *Chthamalus,* even when given the opportunity to do so by clearing the rock of all adult barnacles. What prevented the *Chthalamus* species occupying the whole of the barnacle tidal zonation was that, despite colonization of the lower zone by juvenile forms, these never became established there as adults. The more vigorously growing *Balanus* juveniles either overgrew them, or dislodged them by their greater growth (Fig. 5-5).

These experimental observations by Connell established that the maintenance of individual zones of the species of these two barnacle genera resulted from natural *interspecific competition*.

Competition among Natural Plant Populations

Harper and his co-workers have also established how competition operates between *plant* populations under natural conditions, where it is sometimes a question of competition for soil microsites (1961). Their experiments suggest that the seeds of plants have very precise microenvironmental requirements for germination. A natural soil surface offers a wide range of such microenvironments, which will include a number of "safe" germination sites for each competing plant species in the area. The number of individual plants of each species which finally germinate and become established is correlated with the number of "safe" microsites suitable to each. Harper and his colleagues concluded that there are minute differences in the shape and the surface contouring of seeds. Using quite recent techniques like the scanning electron microscope such differences can indeed now be detected. These minute morphological differences interacted with correspondingly small variations in soil surface

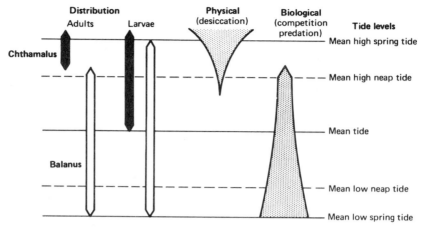

Fig. 5-5. Competition between populations of two barnacle species in the inter-
tidal zone. Young individuals of *Chthamalus* become generally established over
a wide zonal range, but are competitively excluded by *Balanus* in the zone
suitable for its establishment. The faster growing individuals of this second
species push off or overgrow individuals of the other more tolerant barnacle
species. (Redrawn based on several representations of Connell, 1961.)

texture and structure to differentially effect germination and growth of
seedlings of competing species.

From such considerations of the interrelationships between populations in soil communities, there has developed in the last few years a
gradual realization of the significance of secondary plant substances in
relation to population interactions, including the selective type of interaction which has come to be called *coevolution*.

Secondary Plant Substances

Considerable controversy has developed from the theoretical consideration of how secondary plant substances have evolved and whether
these were primarily waste products which have been selected for various
antibiotic activities, or whether particular substances having antibiotic
properties were evolved by adaptation of other naturally occurring
chemicals. Individuals holding the first view maintain that the growth
of all populations is associated sooner or later with an accumulation of
waste products at toxic levels. They consider the antibiotic activities of
certain of these substances have been selectively increased. Individuals
of the other school of thought hold that antibiotic activities are the
direct result of selection processes which develop from interaction at
various interfaces within ecosystems.

Antibiotics

Historically, clinical aspects of antibiotic growth substances were the first of the consequences of chemical population interactions to be extensively explored and utilized. As the search for such antibiotic properties immediately preceded the onset of World War II, with its urgent and extensive need to treat both wounds and epidemics on a vast scale, this antibiotic aspect of interaction attracted a perhaps exaggerated attention. Antibiotics in the clinical sense now most commonly implied by this term are chemical substances secreted by various organisms, and in particular by microbes. They have a regulatory significance in regard to the population dynamics of other species, again especially other microbes.

While the first such antibiotic to be discovered was *penicillin* this was soon followed by *streptomycin,* then by a whole series of others such as *aureomycin,* and *chloromycetin.* Unfortunately the use of antibiotics was initially extensively abused, and antibiotic-resistant strains of many human pathogens developed as a consequence. This tendency was further aggravated with the discovery that certain antibiotics produced increased yields of farm products when introduced into animal feeds. It is known that the genetical factors for resistance which are selected when antibiotic resistant strains are evolved can be passed over to other populations directly by gene exchange. Genetical resistance can be transferred from one population to another even though no mutant allele conferring such antibiotic resistance was originally present in a particular population. In many countries the use of antibiotics is therefore now much more closely regulated than formerly.

Allelochemistry

Antibiotics are only one form of expression of a more general relationship which has come to be known as *allelochemistry.* This is the name given to the interaction between certain secondary plant substances—that is chemical secretions released by the members of one particular population, on the growth of members of other populations at the same or a different trophic level. When it effects competition between species at the same trophic level, it is then generally known as *allelopathy.* Allelochemical interactions can also influence relationships between populations at one trophic level and the next one above it.

One of the most striking demonstrations of allelopathy was that first explained by Muller (1966) in California chaparral. It had been known for many years that chaparral plants have a characteristic pungent and aromatic odor, and that this is due to the presence of volatile substances

Fig. 5-6. Allelochemical effects are commonly observed in chaparral in Southern California and in many other biomes. In the example shown here, terpenes given off in gaseous form from the leaves of sage plants (*Salvia leucophylla*) have suppressed the germination of grass and other herbaceous seedlings in an area about one meter wide stretching out from clumps of sage plants. The cage in the foreground is part of an experiment to demonstrate that rodents are not solely responsible for this effect.

released from the shoots of such plants and known as *terpenes*. What had not been appreciated before Muller's work was that terpenes, when they are adsorbed onto surrounding soil surfaces, suppress the growth of the seedlings of many species in the area. Muller showed that in addition to terpenes, chaparral plants may contain other substances, for example phenolic compounds, which are not volatilized but which are partially water-soluble, and which when dissolved from leaf and branch litter had similar depressing effects on competing species (Fig. 5-6).

This whole allelopathic system of terpenes, phenolic compounds, and other substances is related to the periodic fires that occur in chaparral, during which all plant debris on the soil surface is destroyed and any substances adsorbed on this surface burnt off. Many so-called "fire succession" species are therefore held dormant as seeds in the soil until this periodic occurrence of fire. Then they are no longer suppressed by the various allelopathic substances, and germinate to provide the pioneer species colonizing the burned areas.

Many other forms of allelochemical reactions are known. It seems that in general the volatile terpenes occur in more arid climates, and water-soluble phenolics in areas of higher rainfall. Grasses are known to secrete allelochemical substances, some of which suppress such vital soil processes as nitrification and ammonification (Whittaker et al., 1971). Even peach orchards contain an alkaloid *amygdaline* which has been secreted by peach trees, and behaves as an allelochemical substance. There is indeed believed to be an extensive series of such allelochemical interactions between producer-consumer and consumer-consumed populations at various trophic levels. From an applied standpoint the most intriguing class of substances are the juvenile hormones, including *ecdysone,* which control the metamorphosis of an insect pupa to an adult. They or their analogues commonly occur also in plants, and the immunity from insect grazing they must confer should obviously be capable of exploitation in the search for better and more selective pesticides (Williams, 1967; Miller, 1969).

Chemical Interactions Between Populations

This kind of chemical interaction between populations is being increasingly stressed in studies of coevolution (Ehrlich and Raven, 1965). It has also been the subject of some very elegant and fascinating experimental investigations like those of Brower (1969). Such considerations are of central interest in regard to ecosystem evolution and are comparable in importance with those of energy and nutrient circulation. Indeed, without the existence of such chemical interactions between populations at different trophic levels it is difficult to conceive of any stable trophodynamic relationships. Had plant species not frequently been unacceptable as food for one allelochemical reason or another, insects would long ago have wiped them out entirely. Similarly, predators such as birds in their turn might long ago have annihilated all insect species. Perhaps even some predator or predators unknown might have pushed *Australopithecus* to extinction, and so have prevented our own species from developing—to later devastate the earth, it would seem.

A consideration of secondary growth substances is therefore not only of great fundamental significance, but also of immense importance in applied ecology.

A more general observation on populations and communities in relation to competition was made by Charles Elton (1946) and this leads into a further aspect of biotic influences in evolution which must be considered. Elton comments that individual genera in both producer and consumer populations of ecosystems are generally represented by a single species. He notes that it is unusual for species so closely related taxonomically as to permit their inclusion in the same genus to coexist in the same trophic level, and where they do it is usually because of the occurrence of what has come to be known as *character displacement.*

Character Displacement

The phenomenon of character displacement can occur when two otherwise allopatric populations become sympatric in a portion of their dispersal area. Individuals from each species in the sympatric section of the two populations exhibit a greater divergence from one another than they do in the allopatric section (Brown and Wilson, 1956). This phenomenon, illustrated in Fig. 5-7, may be so intense as to displace one of the competing species into a neighboring trophic level, in which case it is known as *trophic displacement*. Schaffer (1969), suggests that trophic displacement occurred in the early history of the genus *Homo*, when some three or four million years ago two forms of terrestrial social anthropoids with hominid characteristics were evolving in the same area of tropical Africa. Trophic displacement caused one of these forms to evolve toward a greater tool utilization and tool manufacture, emphasizing a carnivorous diet, while the other form became a specialized vegetarian, thus remaining at the herbivore trophic level (Fig. 5-8).

The operation of these various factors governing competition between species which have been examined in this chapter result either in the extinction of particular species or in the separation of the range of variation of species populations throughout the trophic levels of particular ecosystems. Each species in either case therefore comes to be associated with a particular and unique set of biotic and abiotic influences which have become known as the *niche characteristics*.

The Niche Concept

The term *niche* has been used in ecology with a variety of meanings. In the sense of an *ecological niche*, first introduced by G. E. Hutchinson in 1951, it is intended to refer to the totality of biotic and abiotic factors to which a given species is uniquely adapted. Hutchinson conceived this adaptation as locating the species at a given position in a multidimensional space delimited by the parameters of these operative habitat factors.

The term niche is sometimes given a special meaning, as when an island is said to have a number of *empty niches*. In its strict ecological sense there can be no such thing as an empty niche, because an ecological niche is an attribute of a species. What the term means when used in this special sense is that the habitats of the island ecosystem include a number of resources not as yet being fully exploited by the island communities. To avoid the possibility of confusion in meaning, ecologists have available two other terms referring to spatial and environmental

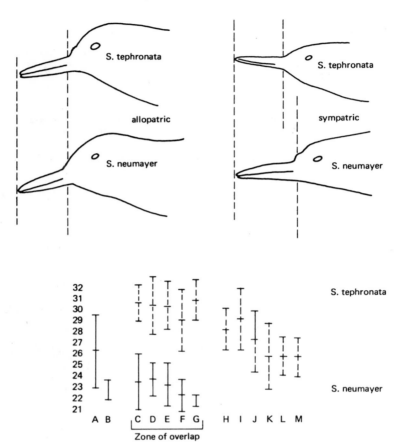

Fig. 5-7. Character displacement in Asiatic nuthatches illustrating geographic variations in the bill length and facial stripe in two species, *Sitta neumayer* and *S. tephronota*. Observations were recorded in A, Dalmatia and Greece; B, Asia Minor; C, Azerbaijan and Iran; D, Kermanshah; E, Luristan and Bakhtiari; F, Fars; G, Kirman; H, Baluchistan; I, Southern Afghanistan; J, Khorasan; L, Northeast Afghanistan; M, Ferghana and Tian Shan. In A and B where *S. neumayer* exists alone, and H–M where *S. tephronota* is the sole species, bill length and facial stripe dimensions are quite similar. In C–G, where the two species overlap, measurements indicate these two characters differ appreciably as between one species and the other. (Redrawn from Vaurie, 1950; reproduced with permission of the publisher.)

features, *area* and *habitat*. The *area* of a species or any other taxon refers to the total extent of its geographical range of dispersal; this can be plotted on a map. The *habitat* of a species describes in a single word or phrase the totality of abiotic factors to which the species is exposed in this area. Thus we can talk of marine habitats, coastal habitats, marsh habitats, disturbed habitats or even dry habitats.

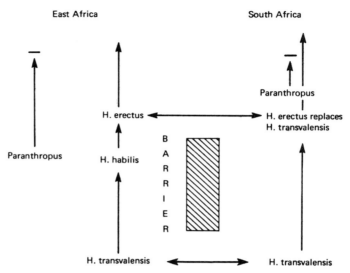

Fig. 5-8. Character displacement in early man, illustrating one of the theories presented to account for the emergence of species of the genus *Homo* from the hominid complex which included such forms as *Paranthropus*. This theory supposes that competition between the several sympatric forms of originally omnivorous hominids caused one group (*Homo erectus*) to become basically carnivorous, while the other (*Paranthropus*) adopted an essentially vegetarian diet. Because *H. erectus* became a secondary consumer and *Paranthropus* a primary consumer, this adaptation can be called *trophic displacement*. (Redrawn based on Schaffer, 1968.)

Niche Differentiation

Because of the competitive exclusion principle described earlier in this chapter, if two or more otherwise competing species are to occur sympatrically in the same ecosystem, there has to be *niche differentiation*. That is, there must be a shift in the individual species requirements which minimizes competitive interactions; as noted, one way this is achieved is by character displacement. Competition may be lessened in a number of other ways, although these could usually be included within a broad interpretation of the meaning of character displacement. One of the commonest is variation either in the nature or size of the preferred food. For example in parts of Africa such as Natal, two species of rhinoceros, the black and the white, are sympatric. The black rhinoceros is, however, a *browser*, the white rhinoceros a *grazer*. One eats woody plants, the other grasses and herbs, so that they are not in competition for the same food resource. In tropical rain forest some species of bat may be insectivorous and feed on adult mosquitoes, so may species of insectivorous birds and these therefore occur at the same carnivorous trophic level of the same ecosystem. The bat species however are nocturnal, the bird species diurnal, and the two do not therefore compete.

The seasonal development of plants within a temperate woodland such as that of the northeastern United States provides another example of niche diversification. Three basic forms of plants may be recognized in these essentially deciduous forests, herbs, shrubs, and trees. The forest herbs such as species of the genera *Trillium, Asarum* and *Viola,* begin their seasonal vegetative growth before even all the snow has cleared from the ground, and complete the major part of their reproductive cycle before leaves have appeared on either the shrubs or the trees. They are therefore able to photosynthesize using the major part of the incident radiation of the forest which is as yet interrupted only by the bare tree and shrub branches. The leaves on the shrubs, which are species of such genera as *Ribes* and *Viburnum,* likewise develop comparatively early, and again initially enjoy largely uninterrupted light. Finally, in late June or early July, the leaves of the oaks, hickories, maples, beeches, and other deciduous trees expand, thus intercepting the light and greatly reducing the amount of incident radiation which the two subordinate life forms of the forest receive. Competition for light between the three basic producer life forms is thus partially avoided by this phenological staggering to achieve a maximum vegetative and reproductive activity in the community.

Niche diversification and some other aspects of the niche concept are considered further in the final chapter of this text dealing with communities. For the remainder of this chapter it is necessary to examine certain other implications of the interaction between species besides the essentially evolutionary ones which have so far been considered. While these evolutionary aspects could be regarded as the type of competition described by Park under the heading of *exploitation,* that is competition for the same resource, the kind of competition now to be considered falls into his second type, which he describes as *interference,* one aspect of which—allelochemistry—has already been discussed.

Interference Between Populations

The nature of *interference* between populations has been quite extensively examined. Slobodkin (1962) for example finds this referable to one of five types which are as follows:

1. Two populations competing for the same resource of the habitat (Park's *exploitation*).
2. A second population serving as a resource for the first.

3. The first population serving as a resource for the second.
4. The two populations being of mutual benefit.
5. The two populations being quite independent (no interference).

This basic statement on the nature of interference has been more elaborately described in a table prepared by Odum (1959) which sets forth the variation definable in such interactions (Table 5-1). Both these

TABLE 5-1

Degrees of interference between two populations A and B—partly based on a scheme first presented by Odum, 1959.

Nature of Interference	Effect of Interference		Result of Interference
	Population A	Population B	
Neutralism (no interference)	nil	nil	none
Mutualism (symbiosis)			
a) Protocoperation	favorable	favorable	favorable but facultative for A and B
b) Commensalism	favorable	nil	favorable and obligatory for A only
c) Amensalism	unfavorable	favorable	unfavorable and unnecessary for A, favorable and facultative for B.
Dependence			
a) Parasitism	unfavorable	favorable	unfavorable and unnecessary for A, favorable and obligatory for B.
b) Predation	unfavorable	favorable	unfavorable and unnecessary for A, favorable and obligatory for B.
Competition	unfavorable	unfavorable	unfavorable and unnecessary for both A and B, one of which will eventually be eliminated.

schemes require a consideration of the general ecological phenomenon commonly described as *symbiosis.*

Symbiosis

The original and most limited interpretation of symbiosis included only a few special relationships such as that between the algal and fungal components of lichens, and that of the microbial populations of nodule bacteria and their leguminous hosts. The broader concept of symbiosis made possible when it is considered as in Odum's scheme as an expression of *interference* between competing populations, includes not only these types of relationships but also much broader mutualistic ones previously described as coexistence, commensualism and a variety of other terms. This broader view of symbiosis can even be extended to cover host-parasite relationships and predation, for these are as much forms of interference between populations as are other consumer-consumed interactions.

It is impossible in this short introductory text to indicate the immense significance of these various widely differing reactions in applied ecology, although some were discussed in Chapter 3 in connection with population regulation. Quite obviously in a world which has finite supplies of fertilizer, the activities of nitrogen fixing organisms, whether of the symbiotic leguminous kind or the several other known types is of great importance. This was mentioned in Chapter 2. Equally significantly, there are many mutualistic interactions which occur in the ecosystems we occupy, which we are unconsciously or deliberately destroying. Many of the forecasts of ecocatastrophies are now based on the anticipated results of this destruction of existing mutualistic relationships (de Bell, 1970). Other examples are described in various volumes in this series.

Bibliography

References

Brower, L. P., "Ecological Chemistry," *Scientific American* 220(2):22–29, 1969.

Brown, W. L. and E. O. Wilson, "Character Displacement," *Systematic Zoology* 5:49–64, 1956.

Connell, J. H., "The Influence of Interspecific Competition and Other Factors on the Distribution of the Barnacle *Chthamalus stellatus*," *Ecology* 42:710–723, 1961.

Ehrlich, P. R., and P. H. Raven, "Butterflies and Plants: a Study in Coevolution," *Evolution* 18:586–608, 1965.

Elton, C. S., "Competition and the Structure of Ecological Communities," *Journal of Animal Ecology* 15:54–68, 1946.

Errington, P. L., "The Phenomenon of Predation," *American Scientist* 51:180–192, 1964.

Griffith, K. J., and C. S. Holling, "A Competitive Submodel for Parasites and Predators," *Can. Entomologist* 101:785–818, 1969.

Hairston, N. G., F. E. Smith, and L. B. Slobdokin, "Community Structure, Population Control, and Competition," *American Naturalist* 94:421–425, 1960.

Hardin, G., "The Competitive Exclusion Principle," *Science* 131:1292–1297, 1960.

Harper, J. L., "Establishment, Aggression, and Cohabitation in Weedy Species," in H. G. Baker and G. L. Stebbins (eds.), *The Genetics of Colonizing Species.* New York: Academic Press, 1965, pp. 243–265.

Harper, J. L., J. N. Clatworthy, I. H. McNaughton, and G. R. Sagar, "The Evolution and Ecology of Closely Related Species Living in the Same Area," *Evolution* 15:209–227, 1961.

Holling, C. S., "The Functional Response of Invertebrate Predators to Prey Density," *Mem. Entomol. Soc. Canada* 48:1–86, 1966.

Lamb, I. M., "Lichens," *Scientific American* 20(4):144–156, 1959.

Limbaugh, C., "Cleaning Symbiosis," *Scientific American* 205(2):42–49, 1961.

MacArthur, R. H., "Population Ecology of Some Warblers of Northeastern Coniferous Forests," *Ecology* 39:559–619, 1958.

Miller R. S., "Pattern and Process Competition," in *Advances in Ecological Research* 4:1–74, 1967.

———, "Competition and Species Diversity," in *Diversity and Stability in Ecological Systems,* Brookhaven Symposia in Biol. No. 22:63–70, 1969.

Muller, C. H., "The Role of Chemical Inhibition (Allelopathy) in Vegetational Composition," *Bulletin of the Torrey Botanical Club* 93:332–251, 1966.

Park, T., "Beetles, Competition and Populations," *Science* 138:1369–1375, 1962.

Vaurie, C., "Adaptive Differences between Two Sympatric Species of Nuthatches (*Sitta*)," *Proc. X Int. Orn. Cong.* 1950:163–166, 1951.

Whittaker, R. H., and P. R. Feeney, "Allelochemics: Chemical Interactions Between Species," *Science* 171:757–770, 1971.

Further Readings in Human Ecology

Cantlon, J. E., "The Stability of Natural Populations and Their Sensitivity to Technology," in *Diversity and Stability in Ecological Systems,* Brookhaven Symposia in Biology No. 22:197–205, 1969.

De Bell, G., *The Environmental Handbook.* New York: Ballantine Books, 1970.

Ewart, W. H., and P. Debach, "DDT for Control of Citrus Thrips and Critricola Scale," *California Citrog.* 32:242–245, 1947.

Holling, C. S., "Stability in Ecological and Social Systems," in *Diversity and Stability in Ecological Systems,* Brookhaven Symposia in Biol. No. 22:128–141, 1969.

Messenger, P. S., "Utilization of Native Natural Enemies in Integrated Control," *Annals of Applied Biology* 56:328–330, 1965.

Schaffer, W. M., "Character Displacement and the Evolution of the Hominidae," *American Naturalist* 102:559–571, 1968.

Schinger, E. I., and E. J. Deitrick, "Biological Control of Insects Aided by Strip-Farming Alfalfa in Experimental Program," *California Agriculture* 14:8–9, 1960.

Williams, C. M., "Third-Generation Pesticides," *Scientific American* 217(1):13–17, 1967.

Wurster, C. F., "DDT Reduces Photosynthesis in Marine Phytoplankton," *Science* 159:1474–1475, 1968.

Review Questions

1. What do you understand by the term *competition*? Provide examples to illustrate your answer and discuss these within the terms of your definition.

2. Explain current interpretations of the term *symbiosis*. Illustrate your answer by discussing some specific examples of the kinds of ecological situations you include within your definition.

3. Describe what is meant by *mutualism*. Discuss how relevant mutualistic associations are with reference to ecosystem function and evolution.

4. Explain the ecological nature of the phenomena included in the term *allelochemistry*. In what way is this a more embracing term than *allelopathy*? What are the major influences of allelochemistry in general ecology?

5. Describe the kinds of ecological interactions which occur between producer and consumer organisms in ecosystems.

6. Discuss predator-prey interactions.

7. What is meant by the competitive exclusion principle? Describe some examples of the kind of interaction which you would include in this term.

8. What do you understand by "character displacement"? Provide and describe examples of this with which you are familiar.

6

Population Behavior

So far in this text we have considered the nature and functions of ecosystems, and environmental influences on these, together with the taxonomic, dynamic, and interactive aspects of the populations of which their communities are composed. One further fundamental aspect of populations must be considered before proceeding to review the nature and functions of these communities, *population behavior*.

Types of Behavior

Behavior has come to have a special meaning in ecology, for which the term *ethology* is sometimes employed, although this word has other connotations. In popular language *behavior* may be defined simply as what an organism does, and basically this is its ecological meaning. The nature of the behavioral activities of organisms based on this definition may for convenience be grouped into three kinds. The first relates to the *accommodation* of the individual population within the habitats of the ecosystems in which it occurs. The second concerns the *regulation* of individual interactions within the population, its social function. The third group relates to the *maintenance* of populations, reproductive behavior.

The first kind of behavioral pattern varies greatly according to the nature of the organism, and its habitat. Terrestrial plants, for example, have very limited mobility, generally restricted to movement of individual organs within a very limited range (Fig. 6-1). Aquatic plants on the other hand may have a quite extensive mobility of either a passive or an active kind (Fig. 6-2). Microbial populations in an aquatic environment may likewise be passively mobile, as they may be in an aerial one, or actively so in an aquatic one. Characteristically, however, whether in a terrestrial, aquatic, or aerial environment the most mobile forms of organisms are animals (Fig. 6-2).

Fig. 6-1. Plant tropisms. One major and very conspicuous difference in behavior between plants and animals is that multicellular terrestrial plants rarely *move* as a whole. Movement is usually restricted to individual portions, and is known as *tropism*. So-called "sleep movements" are one such example. More rapid movement is shown by the shoots of certain plants like *Mimosa pudica* as a response to tactile stimuli,—that is, being touched. The folding of the leaves in *Mimosa pudica,* a so-called "sensitive plant" occurs within several seconds of its being touched.

Because of this characteristic of high mobility that animals possess, behavior has sometimes been regarded as a *capacity for mobility,* a narrower definition than that which we have taken here.

Reproductive behavior has also been especially related to mobility, the term *male* being applied to the moving structure or organism. Social behavior can be regarded as an extension of reproductive behavior, for it favors the survival of a group and therefore the reproduction of the individuals it includes. Selection pressures have tended to operate so as to favor populations that exhibit social behavior. Although in the earlier phases of evolution the relatively simple forms which had been evolved possessed individual genes to deal with individual aspects of behavior, higher forms of animals elaborated a particular form of "nongenetical" behavior to which we generally apply the term *learning.*

Social learning dispensed with the need to evolve individual genes to control individual behavior. Much behavioral response could be *learned,* provided that the organism contained a central nervous system of sufficient sophistication to store the necessary information on stimuli and responses. For the vast majority of animals, social learning provides an ability to react appropriately to particular variations in the biotic or

Fig. 6-2. Taxis. Aquatic unicellular plants commonly possess characteristics which slow their rate of sinking; these include a flattened shape, projections, and oil or gas droplets in the protoplasm. Additionally some are capable of active movement or *taxis* by means of cilia and flagellae, or directed mucilaginous or water streams. Movement in relation to light is shown by most communities of free unicellular aquatic plants, collectively known as phytoplankton. Among nonmotile forms it results from the added flotation provided by the release of gaseous oxygen during *photosynthesis*.

Free-floating microscopic animals (zooplankton) usually display a diurnal depth migration as illustrated here. This taxis is also a response to light fluctuations. (Redrawn based on Wells, 1960.)

abiotic element of the ecosystem in which its population occurs, which have already been experienced. For one group of animals uniquely, members of the genus *Homo* of which our own species *Homo sapiens* is presently the sole survivor, social learning not only provides this ability, but enables individuals to cope with the antithesis—that is, to modify abiotic and biotic elements of the ecosystem so as to make *them* conform to its own behavior patterns.

This learning aspect of social behavior and especially that which occurs in animal populations is what we can term *adjustment*. It returns us to a consideration of the first type of behavioral pattern as defined above, that which is concerned with the *accommodation* of the species within the biotic and abiotic parameters of its ecological niche.

It should be noted that these various aspects of behavior are those with which ecologists are immediately concerned. Behaviorists in this country presently tend to regard their main interests as focused on *learning* and *memory*, and with certain other aspects of behavior considered here as peripheral. Even in Europe, where ethological work has received a greater emphasis, concentration has often been on the manner in which behavioral responses function in the individual organisms rather than in the population.

Adjustment

Organisms adjust behaviorally in two ways, by making internal changes in their organization, or by transporting themselves to a different situation. Either way they need sensory equipment to detect the presence and levels of the substances or factors to which they must respond. This response to information obtained through this sensory equipment is known as *irritability*. We will consider this phenomenon further in the particular context of animals.

Sensory Organs

Most animals have sensory organs that respond to electromagnetic or other type of waves of the forms we call heat, light, and sound. In addition, specialized behavior often results from the development of further sensory organs with or without these first three kinds, which permit the reception of tactile and chemical stimuli. Higher forms of animals, such as our own species, can commonly receive stimuli of all five of these general forms. Our eyes record the patterns of light over a particular range of wavelength, our ears receive sound waves of given frequencies, glands in our skin permit the detection of the temperature of substances, others can register contact with objects and the nature of the texture of their surfaces. Our olfactory organs provide information of the presence of chemical substances in the air surrounding us, and glands in our tongues detect a limited range of dissolved substances placed in contact with them.

We respond to other factors, of which the best known is gravity. Irritability in respect of this automatically sensed stimulus enables us to maintain a particular body position relative to the surface of the earth. What happens when response to the stimulus of gravity is diminished by damage to the inner ear or by the reduction of the capacity to adjust to it as in an inebriate, may readily be observed. Other animals may respond to additional abiotic features of their habitats such as atmospheric or water pressure and electric fields, including perhaps magnetic fields.

Information Processing

The stimuli received from these various sensory organs convey information which has to be transmitted for sorting, interpretation, and possible storage and action. These further functions in higher animals are undertaken by the specialized portion of the central nervous system known as the *brain*. This, or some other portion of the nervous system, is also concerned with the same kind of coordination of information as to *internal* as well as external conditions. It is with the particular involvement of the brain in learning and memory that psychobiologists in this country have been especially concerned as already noted.

Movement

Ecologically speaking the most extensive use to which an animal puts its sensory and information processing equipment is in *movement*. Mobility in an animal is necessary in order to obtain food, to avoid being used as food by a predator, or as part of the reproductive process. Mobility enables an animal to locate itself within an ecosystem in such a way as to achieve the maximum probability of the occurrence of a particular desired result, whether this is feeding, escaping, or reproducing. Animal behavior can thus be regarded as the complex of procedures for achieving simultaneously an appropriate physiological condition and a suitable positioning within a habitat at which the probability of attaining a desired end is highest.

Where this goal favors survival in an *individual* sense, either temporarily or reproductively, the behavior is sometimes described as *egocentric*. When the behavior favors this in the *social group,* but not necessarily for the individual who is a member of it, the behavior has been called *altruistic.* These are two very convenient terms to use, and avoid some of the difficulties encountered when employing the term "group" as in "group selection."

Egocentric Behavior

Under this heading we will review some of the characteristics of behavior which locate an animal in a particular condition in a given ecosystem. We are not concerned for the moment with altruistic behavior, which is considered when discussing reproductive procedures.

Habitat Selection

The location of an animal within its habitat range in a particular ecosystem is believed to be basically a response to genetically determined characteristics. Much evidence for this supposition is drawn from experimental work on deer mice (*Peromyscus* spp.). Harris (1952), for example, showed that two subspecies of P. *maniculatus,* the one normally encountered in grassland, the other in woody vegetation, would select the appropriate simulated artificial habitat, even when individuals born and raised under laboratory conditions and not previously exposed to such a choice were used.

Wecker (1963) pursued this problem of habitat selection further in a series of experiments comparing the behavior of wild-trapped and laboratory individuals of a subspecies of P. *maniculatus.* He constructed a large rectangular enclosure (Fig. 6-3) 1,600 square foot, and placed this with its

Fig. 6-3. Habitat selection in deermice. Photographs of the enclosures constructed to test the reaction of animals from a subspecies of deermouse (*Peromyscus maniculatus*) to different types of cover. In selecting between a "wood" (below) or "field" (above) type, all animals showed a preference for a "field" habitat, whatever their previous experience, thus reflecting the "natural" habitat of the species. (From Wecker, 1963; reproduced with the permission of the publisher.)

long axis lying across a natural wood-field boundary. Animals introduced to the enclosure thus had a choice of a wood or field habitat, and their movements were registered on automatically recording treadles located about the enclosure. Wecker tested the reactions of field-caught mice, their wood-raised offspring, their laboratory-raised offspring, laboratory mice, their field-raised offspring, and their wood-raised offspring. He found that all six groups showed a preference for the field section of the enclosure. Apparently early experience in a wood habitat did not cause any favoring of this habitat against a field one, although some reinforcement of preference for a field habitat could occur as a result of early experience.

Using adult chipping sparrows, Klopfer (1965) demonstrated experimentally their preference for pine as opposed to oak foliage, even when raised in contact with neither form. Such experiments as these of Wecker and Klopfer confirm that over a wide range of animals, habitat selection is basically genetic. However, there is some plasticity in this respect, otherwise animal groups would never extend their habitat range except as the result of mutation. The most dramatic examples of habitat switches may be encountered in adaptations to urban artifacts. Rats have moved into sewers, snakes into house roofs, pigeons onto window ledges, starlings onto statues, bedbugs into bedsteads, fleas into clothes, moths into woolens, borers into furniture, beetles into grain silos. Economically the plasticity in habitat selection which such animals have exhibited when moving into urban ecosystems causes damage running into billions of dollars annually.

Feeding Behavior

The food-seeking behavior of many animals is determined by both innate behavior patterns and by learned reactions. Whatever its source, food-seeking behavior tends to become comparatively stereotyped and a particular chain of reactions is invoked by specific stimuli. An example of a stereotyped reaction invoked in the first instance by essentially genetical characters is provided by the solitary digger wasps (*Sphecidae*).

Among digger wasps, each species relates especially to a quite limited range of prey, commonly a single genus or even a single species, and disregards other forms however abundant. Some prey upon grasshoppers, others on cicadas, spiders, or crickets. The adults feed on naturally occurring sugar solutions which are found on fruit, in flowers, and in the honeydew of insects, and thus behave as primary consumers. The female becomes a predator and a secondary consumer only in order to furnish food for the larvae which develop from the eggs she lays. A general account of the behavior of female wasps in this predatory phase has been provided by Evans (1963).

The predatory behavior is initiated with the completion of a burrow or nest, and the female wasp then moves out in search of the specific prey she will place there. The first stimuli used in identifying the prey are visual; further selection is then made by scent. Provided visual and odor stimuli are favorable, the wasp proceeds to seize the potential prey and the stimuli then become essentially tactile. If the combination of visual, olfactory, and tactile stimuli is acceptable, the stinging behavior which paralyzes the prey is triggered. The stinging pattern is usually related to the motor control centers of the prey in such a way as effectively to immobilize it.

As the prey is sometimes significantly larger than the wasp, transport from the place of capture to the burrow or nest may present some logistic problems. The precise method of transport is usually equally as stereotyped as the method of capture, and still further problems which are involved during the insertion of the prey into the burrow or nest are usually solved by stereotyped behavior.

Such stereotyped feeding behavior as this in the solitary predatory wasps is not restricted to invertebrates or animals with comparatively simple nervous systems, but is also encountered in higher animals (Harrison, 1962). For example, as has been described by Meyerriecks, (1960) herons have equally stereotyped methods of feeding (Fig. 6-4). These behavioral feeding patterns in herons are so precise that niche diversification of feeding behavior permits sympatry of as many as nine species which can range the same feeding grounds without apparently encountering competition of an exclusive nature. Somewhat similarly other bird species may utilize different microenvironments of the same habitat (Fig. 6-5).

The feeding of young of the majority of animals also appears to follow highly stereotyped behavior. Among birds, gaping by the young appears to stimulate the feeding by the adults (Tinbergen, 1960). Although this gaping may at first be initiated by an auditory or tactile stimulus, it later may be invoked by visual stimuli as described by a number of workers. Tinbergen and Perdeck (1950) recorded such a reaction in the herring gull where the gaping response is released by any visual pattern which resembles the red spot on a yellow bill which is the characteristic color of the parent's beak. A stronger gaping response was invoked if this pattern was close, and some object of suitable size to be regarded as food was associated with it.

Among mammals especially, such stereotyped feeding behavior on the part of parents and young progresses into learning patterns. Mammals develop through a juvenile stage during which behavioral patterns in response to particular stimuli are learned at first from parents, then in gregarious forms from the total group. Animals, which like mammals learn feeding behavior during this juvenile stage rather than proceeding along genetically determined stereotyped patterns, usually possess a far greater flexibility in their food preferences.

One very special type of feeding behavior is exhibited in animals such as dolphins and bats which locate their prey by echo sounding. This requires the emission of ultrahigh frequency sound waves and the detection by the auditory apparatus of the returning echo. In addition to facilitating the location of prey, echo location permits the animal to

Fig. 6-4. Stereotyped hunting behavior in egrets and herons. Consistent adherence to different patterns of hunting behavior has separated the ecological niches of these and other egrets and herons. Even when they are sympatric there is a minimum of interspecific competition. A, Green heron; B, Common egret; C, Snowy egret; D, Reddish egret; E, Great blue heron; F, Louisiana heron.

avoid other fixed and moving objects which might damage or hinder it. The efficiency of this detecting mechanism is illustrated by the fact that bats have been recorded as able to complete the detection, location, and seizure of insects within a period of a half-second. A personal demonstration of the speed of reaction may be obtained by trying to knock down a bat in flight with a tennis racket as contrasted with a badminton or squash racket. The sweep of the heavier tennis racket is just sufficiently slower when wielded by most people to make it impossible to hit the bat, which can be done with a lighter racket.

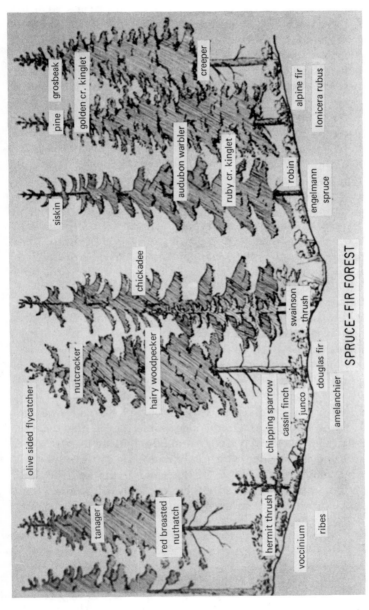

Fig. 6-5. Vertical distribution of territories. In animals like birds and insects whose adjustment is three-dimensional rather than two-dimensional, territories may have depth as well as or rather than width. The example illustrated here indicates the vertical distribution of the territories of some bird species which occupy various horizontal areas of spruce-fir forest in the Northeastern United States. As explained in the next chapter, such niche differentiation is the origin of alpha diversity. (Compiled from various sources.)

Escape and Defense Behavior

In a given ecosystem, top carnivores do not need an alarm system to inform them of the presence of danger from other animal species. All other consumer levels do, but they and top carnivores also will require behavioral mechanisms to detect dangerous or less favorable situations. These could take the form of, say, a forest or grass fire, or the imminence of a severe snowstorm. Detection of such unfavorable conditions is dependent on particular combinations of sensory mechanisms according to the individual species. The recognition of the situation from the information received is usually innate. So also is the escape or defense reaction.

In many animals it is possible to recognize two zones of approach in terms of distance—the *flight distance* and the *charge distance*. Recognition of a potential predator within the flight distance of such an animal will trigger flight. If the approaching animal is not detected until it is within the charge zone, then this will trigger the appropriate charge response. Very interesting experiments are beginning to be conducted on human subjects which suggest that our behavior in respect of escape and defense may also relate to two such definable zones. When approached within the flght distance we tend to edge away or flinch, and within the charge distance we may react positively by pushing, barging, or actual physical exchange of blows. In terms of urban ecosystems and the spatial arrangements within these, it is clearly vital for us to obtain further information on such aspects of human behavior.

Among animals, many of their reactions to danger which invoke flight are contagious. When young are present the behavioral patterns may become aggressive instead of defensive, or may take the form of distraction displays.

One of the main objects of these various behavioral actions is to ensure survival of the individual. At the same time escape and defense mechanisms rarely relate exclusively to one individual. More usually some kind of alarm stimulus is provided which is recognizable as such by other individuals in the community, whether of the same or different populations. Defense and escape behavior can therefore only theoretically be ascribed to distinct categories as either egocentric or altruistic behavior.

Social Behavior Patterns

Within a community, populations at the same trophic level compete with one another to a greater or a lesser degree, as discussed earlier, for some common resource—food, space, or shelter—and any such essentials as are in limiting supply. This competition is not confined to interpopulation interactions, but may occur between individual organisms of the same population for the same resources and for the same reason. We have already mentioned one of the devices that minimizes the adverse effects of such individual competition—the phenomenon known as *territoriality*.

Territoriality

More correctly it should be said that territoriality does not so much reduce competition between individuals as regulate its nature. It is then possible to achieve an optimum population density consistent with the survival of the population from one generation to another at the maximum size, having regard to the resources available.

Territorial phenomena, as already noted, have been observed as occurring at least seasonally among animals as varied as mammals, birds, fish, solitary and social insects, and some reptiles and amphibians (Fig. 3-13). There appear to be three main functions of territoriality—the protection of a nest or den site, the provision of a food supply, and the preempting of a display area. More especially in the course of activities associated with the first two functions, animals tend to move around a specific area of their individual territory which is known as the *home range*. The nest or den site is therefore usually centrally placed within this home range.

The home range like the territory may be large or small, defended or not, or show a seasonal variation in these characteristics. Although not yet extensively investigated in many animals, some interesting data have been obtained from the study of the home range of woodchucks in eastern United States. These demonstrated that while adult woodchucks, both male and female, remain within their particular home ranges, yearling males tend to disperse beyond those in which they were raised (Fig. 6-6). This behavior supposedly is a population-regulating device which has been favorably selected because it prevents overexploitation of the occupied home ranges.

In human societies, studies of territoriality have largely been confined to questions such as the size of the areas occupied by social units of hunting-gathering populations.

Aspects of the territorial requirements of human pair-bond units, like those of the egocentric requirements, are not at all understood, and little work has as yet been done on them. In a world in which people are rapidly losing such attributes as "loyalties" and "social responsibilities," and in which theft of one order of magnitude or another is becoming almost universal, it is clearly urgent for us to know more about our own spatial requirements and any territorial reactions we inherit or learn.

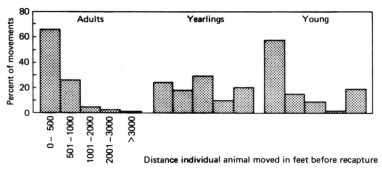

Fig. 6-6. Home-range usually falls within territorial limits if these exist, but it is not necessarily associated with territoriality, as it may not be defended. It can be defined as the area over which an animal will travel in search of food; it may include a more or less centrally positioned "den" or "nest." In the example illustrated, there appears to be some confrontation behavior between adult woodchucks and their offspring which causes the male yearlings as they mature to evacuate the home range of their parents. Supposedly this represents a population regulation mechanism whose selective value is to prevent overexploitation of the home range. (Redrawn: based on several sources.)

Social Rank

Although we have now a considerable number of observations as to social hierarchies in many animal populations, and information on social interaction between these and fertility, population regulation and other behavior, psychologists and ethologists have presently provided few theories as to hierarchies in our human societies. In those animals which exhibit social hierarchies, little energy appears to be expended in maintaining these once relative individual positions have been established. A whole series of often elaborate ritualistic behavior patterns serve both to maintain the hierarchy by threat and to signal recognition of it by deference. Behavioral interactions that occur during establishment and maintenance of dominance are illustrated in Fig. 6-7.

Dominance hierarchies, or social ranks, are most obvious in the social organization of birds. One of the earliest observations was that a flock of domestic fowl possesses a "pecking order" in which the cockerels are usually arranged in order of dominance at the top, the hens in a similar order below (Guhl, 1956). If capons are included they occupy a third and lowest stratum; if the male hormone *testosterone* is injected into either a cock or a hen, it will advance that individual in the social ranking.

While many invertebrates, like crickets, probably exhibit social ranking, few examples have as yet come to light. By contrast, many vertebrate examples are known, social dominance often being positively correlated with size. Fish and reptiles exhibit such dominance hierarchies, as do mammals, where social ranking has especially been studied. The horns

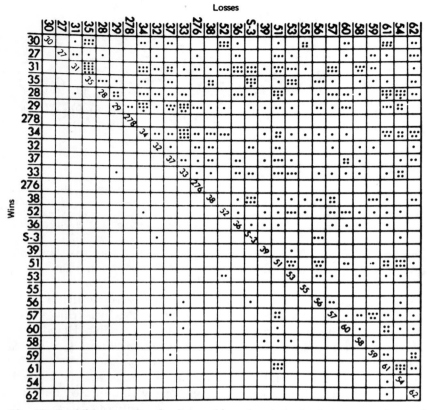

Fig. 6-7. Establishment of a dominance hierarchy. A herd of cows or a drove of steers will establish a dominance hierarchy. Hence the practice of "dehorning" or using "polled" (hornless) breeds, which largely avoids horn damage to the hide arising from dominance struggles. This chart plots the number of encounters necessary to establish such a dominance hierarchy. Vertical rows indicate those encounters which were successful for a particular individual, horizontal ones those in which it was forced to submission. (After Schein and Fohrman, 1965, reproduced with permission of the publisher.)

of many ungulates appear to be related more to the periodic encounters necessary for the establishment of dominance than to defense against predators. Social primate societies show very definite social ranking readily discernible in for example the social organization of a baboon troop on the move (Fig. 6-8).

In such a primate society there are *dominance cliques* rather than individuals at the head of the social ranks. Such an arrangement would appear to have some altruistic selective advantage. It is one or more of the dogs from the dominance clique of a baboon troop which attacks a threatening predator such as a leopard. A social organization which re-

Fig. 6-8. Social organization of an African chacma baboon troop, illustrating the arrangement of dominance hierarchies. Dominant males (A) protect females with young (B). Younger males (C) lead the troop; higher dominance males escort females in oestrus or form the vanguard (D). (Redrawn: based on Washburn and DeVore, 1961.)

lied on one individual alone to perform this altruistic protective function might be thrown into confusion by the death or immobilization of its single leader.

As the top females in a baboon troop are also organized into a dominance clique, offspring born to them enjoy a better hierarchial status than those of individuals further down the social ranking. Hereditary-dominance cliques tend to form in such social organizations. This seems to parallel what may frequently be observed in human societies.

Insofar as dominance and dominance hierarchies have been studied in our own societies, these are patently extremely complicated and frequently become confused. They tend to change according to the nature of a particular activity. The laws and practices of most human societies now have an egalitarian basis in theory, even if this sometimes falls short in practice. This tends to minimize the advantages which progeny of high-ranking individuals enjoy in populations with a less complicated social organization. The more diversified our societies become, the more numerous the various ranking orders. This tends to reduce even further one of the primary advantages which high rank confers on a dominant individual in many social species, that is a mating priority. Such considerations bring us to the third category of behavioral reactions we are examining here, that of *reproductive behavior*.

Reproductive Behavior

Some aspects of the homeostatic regulation of population growth through behavior have already been considered, in particular the feedback mechanism which through its effects on the endocrine glands reduces natural increase when space becomes a limiting factor. Such feed-back mechanisms operate primarily through stimulation of the *hy-pothalamus,* which causes the pituitary gland to release certain hormones that in turn control the activities of various organs (Fig. 6-9). The most important of these hormones is *adrenocorticotropin* (ATCH). This hor-mone stimulates the cortex of the adrenal gland to produce other

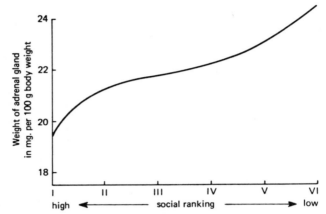

Fig. 6-9. Relation beween social rank and adrenal glands. The adrenals of the dominant mice (I) were approximately the same weight as animals kept in isolation where they were not exposed to social stress. The glands of mice in sub-ordinate ranks became successively larger. (Data extracted from Davis, 1967.)

hormones known as *corticoids.* Together with ATCH, the corticoids reduce the activity of the gonads and thus of the male and female re-productive organs. This results in impotence in the male and infertility and lack of response in the female.

Such hormone controls and behavioral features have been especially studied in laboratory mice populations. The schematic representation of the interactions expressed in one such investigation is illustrated in Fig. 6-10.

Reproductive Behavior Patterns

Behaviorists generally consider that animal behavior has compara-tively less influence on population mortality rate than on birth rates. While admittedly some variants will exhibit behavioral patterns that favor predator avoidance, such effects are of a lesser order of magnitude from those which increase the chance of heterosexual encounters, and the suc-cessful achievement of fertilization. Furthermore, generative processes subsequent to fertilization are commonly long and complicated in ani-

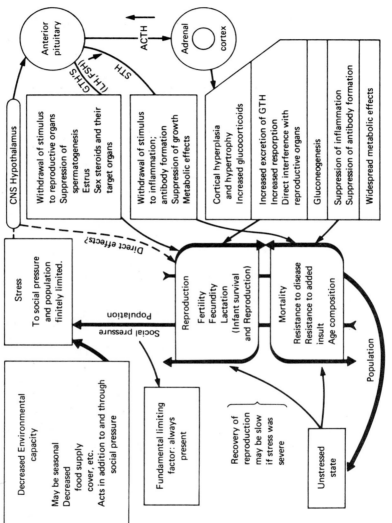

Fig. 6-10. Interaction between hormonal controls and behavior, illustrated from observations on a laboratory population of white mice (*Mus musculus*). This scheme shows the complexity of the interaction in response to social stress, and the results of the influence exerted on the population dynamics of this species. (From Christian, 1959; reproduced with the permission of the publisher.)

mals, providing additional opportunities for them to be still further influenced by behavioral mechanisms. Any such behavior may have regulatory implications with regard to the number of individuals added to the population, and therefore must be included within the parameters expressed by the birth rate.

The first problem which an individual in a heterosexual population faces in reproduction is the identification of other members of its species. The second is the recognition of members of the opposite sex in the appropriate reproductive phase. Recognition patterns employed in overcoming these problems are frequently visual, sometimes chemical or tactile, or a combination of several of these. They are commonly learned in the first few hours of life, as explained later, by *imprinting* (Hess, 1958).

Species-recognition patterns which are visual are, for example, the white tail of a cottontail rabbit. They may be chemical, as the scent gland situated low down on the hind legs of impala. Many others appear to be of an auditory nature, as in frogs, or in birds such as warblers. Recognition of individuals in a suitable reproductive phase involves a type of behavior known as a *courtship display*, which is sometimes extremely elaborate and stylized.

Courtship Displays

Courtship displays have been most extensively studied in birds, because very frequently among this group they have reached an unparalleled level of complication and sophistication. Somewhat curiously, bird species which are generally considered otherwise more primitive, such as penguins and herons, commonly have the most elaborate displays, while species like thrushes and sparrows considered otherwise more advanced, have less complicated ones. There appears to be a correlation between the environment and the principal mode of the display. For example, birds in tropical rain forests where visibility may be limited to several meters may base their courtship display on a series of calls. Predators which search from the sky for their prey generally base their courtship displays on complicated aerobatics. Birds which predate under water may elaborate their swimming performances to courtship displays (Fig. 6-11). In other animals where both movement and appearance may be more difficult to make sufficiently distinctive, tactile stimuli such as nudges, as is commonly the case in deer, eventually convince a heterosexual pair of animals of their appropriately reproductive receptivity.

Courtship displays, as other forms of animal behavior, are essentially under endocrinal hormonal control. Commonly the hormonal activity is seasonal and its level determined by some environmental factor such as the photoperiod (Fig. 2-18).

Fig. 6-11. Courtship displays sometimes assume extremely elaborate behavioral patterns, especially in birds. In some, for example the Flicker (A) and Boattailed grackle (B) the male extends the head upwards, displaying the breast. In others such as the Red-winged Blackbird (C) both males and females display. (Redrawn from various sources.)

Parental Care

The kind of hormonal control of bird behavior which is expressed in courtship displays frequently continues the heterosexual association between mating pairs, or some portion of it, in the form which is commonly known as *parental care*. This may involve protection of a nest site and of eggs located in it by one or both of the parents. The cock ostrich is said to have black plumage because it incubates the eggs by night while the dappled grey female takes her spell in daylight. Or parental care may involve protection of the young by one or both parents, including their feeding in the juvenile stage and their training and instruction at this time.

Perhaps the most significant of all these parental-care patterns among animals is that which conveys information in the form of what we call *learning*. Parental instruction of this nature, together with social learning from other members of the population, is an outstanding characteristic of primate social groups (Rheingold, 1963). It is the "massive" elaboration of these particular behavioral patterns, made possible by effective means of communication, that makes our genus unique among all known living and fossil forms of life.

Both the structure of our human populations and the cities in which they live have to be designed so as to provide for the proper exercise of some form of parental care for our offspring.

Social learning

The response which an individual makes to a particular environmental situation may be entirely unrelated to any previous experience. In the majority of social animals, however, this response is conditioned by stored information as to the results of previous reactions to the same stimuli, by a process psychologists call *instrumental conditioning*. This stored information or experience may have been obtained in one of three ways. It may have come from individual experience, from information conveyed by parents, or from information obtained from the social group.

In more elaborate societies such as a terrestrial primate group, the stored experience has been derived from all three sources. Individual trial and error, or the desire to acquire knowledge in this way, clearly has a high negative selection value. A juvenile that relies solely on personal experience for information that free fall from a 100-foot cliff is fatal is unlikely to pass this characteristic to future generations. Natural selection will favor the survival of genotypes in which the parent is able to communicate such vital information early, and reinforce its memorization, and in which the juvenile is readily willing to receive and retain this information.

The very earliest acquisition of information controlling behavior, particularly that learned by juveniles from parents, as already noted, is called *imprinting*, which especially relates to *following* in precocial birds. For the identification of their own species many animals may rely on imprinting, which is now regarded as a special form of learning that does not operate further once fear of new situations develops. It may be that in humans, which can exhibit fear at birth, imprinting only occurs in the fetus, hence the response to rhythmic sounds resembling maternal heartbeats, and the significance of the "fetal position."

Social learning among juveniles is commonly supplemented or *reinforced* by penalties or awards. When the successful acquisition of new skills or new information is associated with awards it is called *positive reinforcement*. When penalties are incurred, such as slaps or blows, scowling or glaring, this or the withholding of food, is known as *negative reinforcement.* Psychologists and psychobiologists have extensively investigated this process of *instrumental learning.*

In human societies it seems that many adult behavioral patterns can be related to the reinforced learning that occurs in the juvenile stages of life through instrumental conditioning. Some psychologists and sociologists have suggested (Enke, 1969) that an opportunity be taken at this time to inculcate by this means a greater sense of responsibility toward what in known generally as public morality. Of recent years in America, negative reinforcement has tended to be reduced, or even abandoned altogether as a means of ensuring juvenile conditioning. The ecological consequences of this major shift in parental and other social instructional behavior patterns are obviously of considerable interest to human ecologists, but such matters lie outside the scope of both this text and the series to which it relates.

In social animals, selection pressures will favor any behavior pattern that improves the transference of information and its acceptance, as a better-informed animal may be assumed to be "fitter." The extension of this transference to involve the whole social group will clearly also have high selective value. A dead female baboon is unable to convey the information to her offspring that leopards can kill. On the other hand, an inspection by the whole troop of a female baboon which is being eaten by a leopard may permit the information to be conditionally recorded by all juveniles that leopards are to be avoided.

Human societies as already mentioned are unique in two ways. The first is in their huge capacity for communication of information from the present by speech and from the past by writing. The second is the manner in which elaboration of the brain has permitted the acquisition and storage of the social information communicated in this way, and its sorting and retrieval at required times in appropriate forms. Selection of individuals with high capacities for both communication and acquisition of information (which we term learning) must have been subjected to strong selection pressures.

The urban environments in which our species now congregates maximize the opportunities for both these uniquely extensive activities. Hence our apparent compulsive attraction to city life. If we were to devise the means of maximizing communication and learning in some other manner, we would be able if we wished to weaken this tendency to form urban aggregations. Until we do so, it would seem desirable to design urban ecosystems in such a manner as to maximize the opportunities for both communication and learning, and not to try to resist continuing urban concentrations.

This process of learning through communication with other individuals of the species during the juvenile period of development is commonly described as *play.* It provides the opportunity for learning by *instrumental conditioning.* All animals with any kind of social structure indulge in play behavior. With some it is restricted to experiences with siblings and with one or both parents. In more advanced societies it is extended to age-group interchanges, which may be quite extended and quite extensive, and essential for development to normal maturity. Further consideration of such matters would take us too far away from the main purpose of this text and into psychology, psychobiology, and psychiatry.

This all too brief survey of types of behavior patterns completes our review of all ecological aspects of population growth, maintenance and reproduction. There remains for study only the nature, structure, evolution, and function of communities. Such features are considered in the next and final chapter.

Bibliography

References

Alexander, R. D., "Aggressiveness, Territoriality and Sexual Behavior in Field Crickets (Orthoptera: Cryllidae)," *Behavior* 17:130–223, 1961.

Brower, L. P., Brower, J. Van Zandt, and P. F. Cranston, "Courtship Behavior of the Queen Butterfly *Danaus gilippus berenice* (Cramer)," *Zoologica* 50:1–40, 1965.

Christian, J. J., "Phenomena Associated with Population Density," *Proc. Nat. Acad. of Sci.* 47:428–449, 1961.

———, and D. E. Davis, "Endocrines, Behavior and Populations," *Science* 146:1550–1560, 1964.

Davis, D. E., *Integral Animal Behavior.* New York: Macmillan, 1966.

Evans, H. E., "Predatory Wasps," *Scientific American* 208(4):144–154, 1963.

Guhl, A. M., "The Social Order of Chickens," *Scientific American* 194(2):42–46, 1956.

Hamilton, W. J., "Sun-Oriented Display of the Anna's Hummingbird," *Wilson Bulletin,* 77:38–44, 1965.

Harris, V. T., "An Experimental Study of Habitat Selection by Prairie and Forest Races of the Deer-mouse, *Peromyscus maniculatus*," *Contrib. Lab. Vert. Zool.* (University of Michigan), 56:1–53, 1952.

Harrison, J. L., "The Distribution of Feeding Habits among Animals in Tropical Rain Forest," *J. Animal Ecol.* 31:53–63, 1962.

Hess, E. H., "Imprinting in Animals," *Scientific American* 198(3):81–90, 1958.

Klopfer, P. H., (ed.), *Behavioral Ecology,* Los Angeles: Dickenson, 1970.

Klopfer, P. H., and J. P. Hailman, "Habitat Selection in Birds," in D. S. Lehrman et al. (eds.) *Advances in the Study of Behavior* 1:279–303, 1965.

Lorenz, K. Z., *King Solomon's Ring.* New York: Crowell, 1952.

Meyerriecks, A. S., "Comparative Breeding Behavior of Four Species of North American Herons, *Publ. Nuttall Orn. Club* No. 2 (1960).

Rheingold, H., *Maternal Behavior in Mammals.* New York: Wiley, 1963.

Smythe, R. H., *Animal Habits: The Things Animals Do.* Springfield, Ill.: Thomas, 1963.

Thorpe, W. H., *Learning and Instinct in Animals.* rev. ed. London: Methuen, 1963.

Tinbergen, N., *Animal Behavior.* New York: Time-Life Books, 1965.

――――, "The Evolution of Behavior in Gulls," *Scientific American* 203(6):118–130, 1960.

――――, *The Herring Gull's World.* New York: Basic Books, 1960.

――――, and A. C. Perdeck, "On the Stimulus Situation Releasing the Begging Response in the Newly Hatched Herring Gull Chick," *Behaviour* 3:1–38, 1950.

Wecker, S. C., "The Role of Early Experience in Habitat Selection by the Prairie Deer-mouse *Peromyscus maniculatus* bairdi," *Ecological Monographs* 33:307–325, 1963.

Wynne-Edwards, V. C., *Animal Dispersion in Relation to Social Behavior.* New York: Hafner, 1962.

Further Readings in Human Ecology

Birdsell, J. B., "On Population Structure in Generalized Hunting and Collecting Populations," *Evolution* 12:189–205, 1958.

Doxiadis, C. A., "Man's Movement and His City," *Science* 162:326–334, 1968.

Elton, C. S., *The Ecology of Invasions by Animals and Plants.* New York: Wiley, 1958.

Emlen, J. M., "Natural Selection and Human Behavior," *J. Theoretical Biol.* 12:410–418, 1966.

Enke, S., "Birth Control for Economic Development," *Science* 164:798–802, 1969.

Spilhaus, A., "The Experimental City," *Science* 127:9–16, 1958.

Tiger, L., and R. Fox, "The Zoological Perspective in Social Science," *Man* 1:75–81, 1966.

Tinbergen, N., "On War and Peace in Animals and Man," *Science* 160:1411–1418, 1968.

Review Questions

1. Describe what is meant by *imprinting*. What selective advantages could this convey?

2. Discuss the meaning and significance of *social hierarchies*.

3. State the commonly recognized types of *behavior*. What is the ecological significance of each type you define?

4. Discuss the ecological implications of social learning.

5. Describe existing evidence as to the nature of the determining factors which locate a given animal population within its habitat range.

6. Discuss the influence of *intuitive* and *learned* responses in feeding behavior.

7. State, describe and illustrate the main functions of *territoriality*.

8. What ecological significance may be attributed to *courtship displays*?

9. Describe what you understand by the term *instrumental conditioning*. What part may *reinforcement* play in this process?

10. Discuss what is implied in the ethological meaning of the term *adjustment*.

7

The Nature and
Structure of Communities

We have now reviewed all aspects of ecosystem structure, function and development, except the nature of the *communities* which comprise their essential living or biotic element. Communities represent both the best known and the most tangible concept in ecology. The complex of terrestrial communities which surrounds us is one of the most familiar of the living patterns that we recognize. Even for those who are not biologists or ecologists, terms such as *desert, forest, grassland, swamp* conjure up pictures of specific plant and animal communities which we have come to associate through instrumental conditioning with particular habitats. The aggregation of populations into community groups, united in a common activity or encountered in a common habitat, is thus a familiar popular concept, although it is one now increasingly challenged on theoretical grounds (Ehrlich and Holm, 1962).

In earlier portions of this text it has been impossible to consider various aspects of ecology without referring to certain features of the structure and functions of communities. Thus we have already considered such matters as the *niche concept, adaptation, convergent evolution, ecoclines,* and *succession.* Similarly a number of the vital functions of communities have been examined in relation to those of ecosystems, from which they are inseparable. Any reality that an ecosystem possesses derives from the functional relationships between its constituent communities and their environments.

In this final chapter we will therefore reexamine certain aspects of community form and function, and associate these with the remaining characteristics of communities which have yet to be considered.

Characteristics of Communities

In common understanding communities have three essentials. They occupy a definable area—that is, we can describe where to find them, as in *marine communities, urban communities,* or *forest communities.* They are composed of populations; we know for example that a Californian redwood community includes besides redwoods the mule deer, black bears, butterflies, mosses, and all the other forms of life which this expression calls to mind. Thirdly communities have a typical environment; in the case of the redwoods, a cool, moist, still one which we consciously or unconsciously picture as different from say that in which we expect to find stands of saguaro cactus.

As we noted when discussing interactions between populations and their habitats, communities are not invariably discrete units. Commonly their boundaries merge imperceptibly along environmental gradients, forming an ever-changing complex which we defined as an *ecocline.* Comparatively recently an ecological technique known as gradient analysis has been developed for the field investigation of gradient relationships along such ecoclines (Whittaker, 1967).

Gradient Analysis

The current field technique of gradient analysis was elaborated by R. H. Whittaker. He also explained succinctly the range of theoretical possibilities regarding the interrelationships of populations and communities along an ecocline. Following Whittaker there are four such possibilities:

1. Competing dominant species establish sharp boundaries; other species associate with these dominants and are adapted in a noncompeting manner to their various niche characteristics (Fig. 7-1A).
2. Sharp boundaries develop between competing dominants, but no subordinate species cluster about them (Fig. 7-1B).
3. While no sharp boundaries between species develop, there are nevertheless observable population clusters (Fig. 7-1C).
4. As the last with no sharp boundaries, but no observable clusters either (Fig. 7-1D).

Essentially similar theoretical considerations to case (4) were originally developed by Gleason (1939) in his "principle of species individuality." These same conclusions also emerged from the pioneer field studies of the Wisconsin school founded by J. T. Curtis (1959), and extended to a multidimensional concept by others (Buell et al. 1966).

Field observations as to the validity of this fourth, or any other of the several hypotheses, have especially concentrated on altitudinal tran-

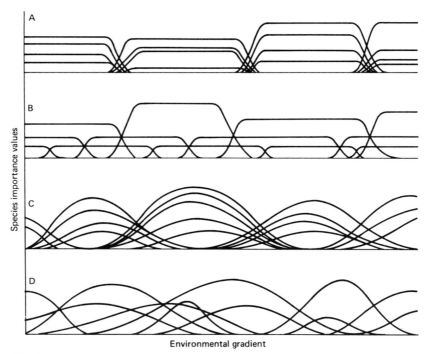

Species importance values

Environmental gradient

Fig. 7-1. Four possible distributional arrangements of species along a gradient. A, dominants with discrete boundaries within which associated species form clusters; B, dominants have discrete boundaries, subordinate species do not cluster; C, no discrete boundaries occur but species tend to cluster; D, no discrete boundaries, no clusters. Field observations suggest arrangement D is the one usually encountered; see text and Fig. 7-2 for further comment. (Redrawn: based on Whittaker, 1970.)

sects. These are prepared by replicating and averaging samples from various heights in which such information as limiting environmental factors and community species composition is recorded. Whittaker's series of such field studies support the last hypothesis, that there are neither sharp boundaries between populations nor associations between them in regard to occurrence and distribution. When frequency of occurrence is plotted against position along such an altitudinal gradient, each species shows a characteristic bell-shaped curve whose peak and base is unique and never coincides with that of another species (Fig. 7-2). This means in effect that there is a *community gradient,* but not a community mosaic. Other field work on community structure also failed to reveal clusters of species associated together to form discrete communities.

An analogy is sometimes drawn between community gradients and the spectrum of visible light. Although the variation in the wave lengths included within the visible range is *continuous,* most people would be

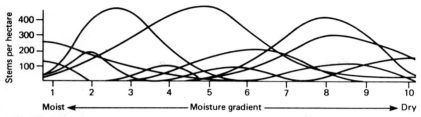

Fig. 7-2. Distribution of tree species along an environmental gradient—data from the Catalina Mountains, Arizona. The species populations are plotted by density of tree stems along the vertical axis, and a moisture gradient on the horizontal axis. There are no sharp boundaries and no clustering of species observable along this moisture gradient. Compare with example D, Fig. 7-1. (After Whittaker 1967; reproduced with permission of the publisher.)

prepared to argue that the spectrum is composed of seven primary colors, red, orange, yellow, green, blue, indigo, and violet. This is indeed the basis for the artist's palette. In the same way the life zones along a mountain gradient in North America first demarcated by Merriam (Fig. 2-9) appear to be real. Actually they are no more substantial than the primary colors, but indicate as they do particular stages in a continuum—in this instance a continuous mountain community gradient.

The reality of community discreteness seems even more plausible following the application of some methods of statistical analysis. In Fig. 7-3 the results of one such ecological exercise are illustrated. The communities demarcated here have been determined by estimating the statistical probability of two species simultaneously occurring in the same field sample area. A very similar field exercise is illustrated in Fig. 7-4.

Ecologists have expended much effort in attempting to substantiate or disprove the community-gradient or community-mosaic concepts. It is very similar to the case of variation within and between populations discussed in Chapter 4. It appears certain that in some instances environmental discontinuities, a mountain chain, a river or a fence line, can insert a discontinuity into community-gradients and thus produce a parallel discontinuity in species representation. In such circumstances there will be community mosaics.

Despite the existence of ecoclines and continuous variation in community composition, it is usually possible to select a set of community features, including the names of the most frequently occurring species of plants and animals, which can be used to identify the commonest general type of community distributed through a particular region. If this is a stable community it is generally known as the *climax community,* and the living complex which it forms can be known as a *biome.*

The term *biome,* introduced earlier in this text, places less emphasis on the nature of the species representation than on the general *structure*

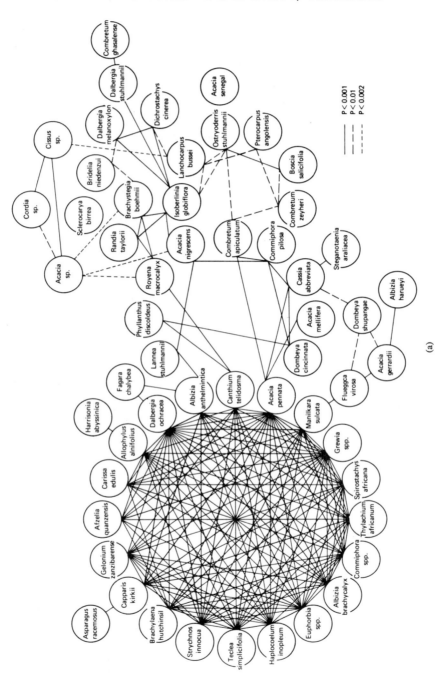

Fig. 7-3. Statistical association between species. (a), the statistical significance of associations between species, determined by a chi-square test, separates this savanna woodland into five communities. On a basis of the probability of their occurring together, clusters of species can be grouped into these five (b). On this statistical basis, hypothesis c in Fig. 7-1 appears to be the best fit accounting for population distribution in this savanna woodland. B, is the schematic summary of the chi-square correlations shown in A. The five species correlations are

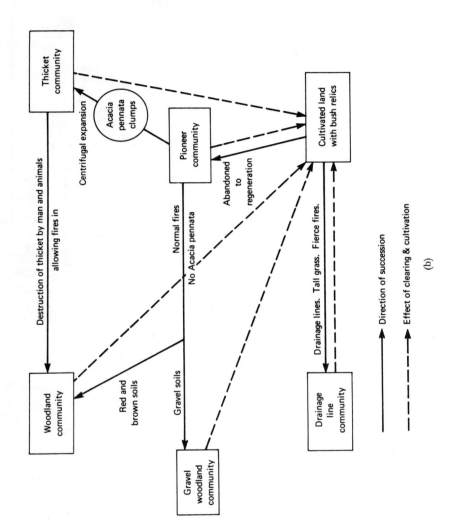

(b)

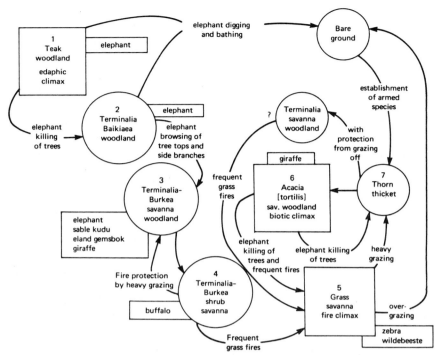

Fig. 7-4. Species clustering in a game ecosystem—a field assessment of associations between plant and animal dominants in various communities of the Wankie Game Reserve in South Central Africa. This apparent clustering was subsequently supported by sampling and statistical analysis similar to that illustrated in Fig. 7-3. In both instances the apparent discontinuities probably arose as a result of the operation of overriding discontinuities in major factors such as fire and grazing. (Redrawn: based on Boughey, 1963.)

of the community. The *grassland biome* of North America for example is characterized by perennial grasses, but the grass species and even the genera represented vary very considerably. On the other hand, the *coniferous biome* would usually include the coniferous trees Douglas fir, lodge pole pine or jack pine, and black bears, racoons, beavers, and deer mice.

Part of the reason for the existence of such diagnostic features is that in a given biome *convergent evolution* has resulted in many populations of widely different taxonomic relationships assuming a generally similar morphological appearance and physiological functioning. Thus in chaparral for example, as was cited in Chapter 4, a species of the genus *Adenostoma* in the family Rosaceae may look very similar to a species of the genus *Erigonum* in the Polygonaceae. Desert succulents as a group typically open their stomata at night instead of in the daytime. This common behavior implies no taxonomic relationship, only an ecological one.

Niche Space and Species Importance

Niche diversification within a given biome results as was discussed in Chapter 4 and again in Chapter 5, in *adaptive radiation* among its populations. Convergent evolution may produce many *ecological equivalents*. There is, however, one further kind of diversity which is of considerable importance in regard to community characteristics.

As previously discussed, this is attributable to the existence of *ecological niches*, the total range of the biotic and abiotic features which characterize individual populations. Some of these niche characteristics are relatively intangible—for example, the level of competition with another species. Others are more readily quantified and thus measurable. So a plant population has for example a particular height range or phenology, an animal population a given density or age structure.

Vertical Structure

Because plant populations have a limited height range, plant communities have a *vertical structure*, whose nature depends upon the height ranges of the different species it includes. In a temperate woodland such as an oak-pine forest this vertical structure may be very simple, as is indicated in Fig. 7-5, because this particular niche characteristic of species height does not greatly vary among the dominant species. That of tropical rain forest, illustrated in the same figure, may be appreciably more complicated, because this community not only has more species but these exhibit a greater diversity in height. In any case the individual plant species have ecological niches whose parameters in terms of vertical height may be measured, recorded, and illustrated.

Such parameters are not necessarily constant, but may change with time. Saplings may grow from one level to another. The vertical distribution of animal populations may or may not coincide with the structural subdivisions in vertical height of the plant populations. In a tropical rain forest the canopies of the over-storey trees may house a bird population whose niche characteristics locate it in this particular zone. A mosquito population on the other hand may at one time of day, or one season of the year, be located in one zone, at another time in a different zone, according to fluctuations in factors such as humidity within the forest.

Representation of Niche Space

Despite such qualifications, we can take two features of a forest such as gradient of height and the extent of foliage persistence (deciduousness) among tree populations, and plot these one against the other. The two-dimensional graph which results illustrates a two-dimensional niche space

A. Deciduous forest profile

B. Moist forest profile

Fig. 7-5. Vertical structure of communities. A represents the layering observable in a pine-oak forest in the northeastern United States. B is the vertical structuring of a tropical rain forest in Central Africa. The amount of alpha diversity possible in B will be far higher than that in A. (Based on drawings from several sources.)

characterizing the tree populations which have become associated in the particular community we are examining. Within this niche space, these populations are distributed according to the competitive exclusion principle. Each species occupies its own particular niche area, which differs from that of other associated species.

If from this two-dimensional illustration we now proceed to develop

Hutchinson's concept of n-dimensional niche hyperspace for each of the species involved, the centers of their population hypervolumes will be located somewhere within each niche hyperspace, and will be separable from the centers for the other populations in the community.

The amount of niche hyperspace occupied in the ecosystem by each plant population is an indication of the *relative importance* of that species in the community. It is the measure of the extent to which it utilizes the resources of the ecosystem available to the community, and the proportion of the productivity of the ecosystem which results from the activity of that particular population in the community. Determination of species *importance values* is basic to the planning of management policies aimed at the preservation of particular ecosystems.

For animal populations, the importance value is related to the *density* of the population, but its best measure is again the proportion of productivity which results from its activities.

Community Species Composition

Concepts and expressions of *relative importance* and *niche characteristics* are closely related to another community characteristic—that is, the number of species populations involved in a particular community. This is known as the *species diversity*; it varies according to the status and nature of the community. Some climax communities may have a minimum diversity. This applies particularly to climax communities in situations where the *predictability* of the environment is low. A botanical inventory of the coniferous forest on the Canadian shield reveals an almost monotonous repetition of 20 to 30 plant species. A similar inventory of a very much smaller area of tropical rain forest would produce a list of 10 or 20 times this number. The corresponding animal and microbial populations would exhibit the same magnitude of variation in number. If we grade terrestrial communities latitudinally, starting from the north polar regions and proceeding to the wet tropics, there is an observable correlation with species diversity.

Latitudinal Species Diversity

The explanation for this latitudinal gradient in species diversity which is most commonly observed is that few species can survive the rigorous environments of higher altitudes. Slobodkin and Sanders (1969) have recently restated that it is a question of the *predictability* of the environment. Where variations in the abiotic element of the ecosystems are minimal, as in the wet tropics, the same range of genetic diversity does not have to be maintained in a population to enable it to survive or at least to enable a portion of its phenotypes to survive irregular and wide fluctuations in limiting environmental factors. In other words, the niche characteristics of species in the tropics may be compressed within much closer limits than those in the arctic. *Species packing* in such predictable environments leads to enormous increases in the species diversity of tropical communities.

Not all workers agree either with this supposition of Slobodkin and Sanders, or that species packing occurs in the tropics. Some would argue that it is a question of evolving forms with sufficient plasticity to survive more widely varying environmental factors, and of adaptation to more critical conditions requiring more specialized features to permit survival.

Dominance

The importance value of a population in a particular community is related to what has already been described as *dominance*. Importance values indicate the contribution of the population to the energy budget of the community within a particular ecosystem; the dominants are the populations through which the major portion of the energy flow proceeds. At the producer trophic level it is usually the tallest plants that are dominants, because it is they that intercept the major portion of the solar radiation energy.

Plant dominants are therefore normally adapted to major characteristics of the macroenvironment. Because of the *convergent evolution* already mentioned they tend to have the same *life form*. The chaparral communities of the Pacific Southwest not only have *Adenostoma* (chamise) and *Eriogonum* (buckwheat) species of comparable life form (Fig. 4-7), but all the chaparral plant dominants are quite similar morphologically and physiologically. The wet-cold-winter, dry-hot-summer characteristics of the macroenvironment of this community favor deep-rooted, relatively low-growing shrubs, with small leathery leaves, capable of restricting water loss during periods of water strain, but equally capable of rapid growth under favorable conditions. Other similarities may be noted, such for example as the fruits commonly being either succulent and bird-distributed, or some kind of nut- and rodent distributed.

It is thus convergent evolution producing life-forms characterizing the dominants of particular communities which permits the creation of *biomes* or *formations* based on these dominants. Other plant species of the community, and most animal species, whether dominants or not, are more closely related to microenvironments than to the macroenvironment, as is even more so the case with the microbial reducer communities.

As was discussed in the preceding chapter, the *mobility* of animal populations permits them frequently to *escape* from adverse conditions of the macroenvironment. Thus in the Californian chaparral out of four species of hummingbird, only one is resident, the other three migrate southward into Mexico during the winter months when flowering, and therefore food supplies, in the form of nectar, are minimal. In winter caribou leave the tundra where they have fed during the arctic summer to winter in the taiga. Many chaparral plants are annuals, permitting them to avoid the dry summer months (August through October). Some chaparral animals hibernate, especially the various species of snake, thus persisting in a dormant condition through the inclement winter months (December through March).

The rodent dominants of the chaparral community all possess seed pouches—pack rats *(Neotoma)*, deer mice *(Peromyscus)*, pocket mice *(Perognathus)*, kangaroo rats *(Diplodomys)*, voles *(Microtus)*, and harvest mice, *(Reithrodontomes)*. With the single exception of the pack rats, they all live in burrows, thus enabling them to avoid exposure to the high summer surface temperatures (Fig. 2-10). This is especially important as they must survive on metabolic water—water they synthesize by oxidation of hydrogen during their metabolic processes. Only occasionally do they utilize the stored water they can obtain by eating succulents, as pack rats do when feeding on prickly pear. The chaparral rodent species are nearly all nocturnal, thereby reducing both heating effects and water loss, and also the level of predation. At the same time this forces their main predators to become nocturnal. The dominant secondary consumers which predate rodents are foxes, coyotes, skunks, bobcats, racoons, opossum, gopher snakes, and owls, all nocturnal predators. A few predators such as hawks and rattlers apparently find a sufficient number of diurnal lagomorphs (e.g. cottontails and jack rabbits), and rodents on the move in daylight, to support their diurnal populations.

In the same way we could list the main features of desert dominant life forms (Whittaker, 1970) or the characteristic life forms of the typical communities of any biome. Each would illustrate a particular set of community adaptations.

Niches and Gradients

Community adaptations, while thus identifying particular plant or animal species with evolutionary trends which result in convergent evolution, nevertheless do not entirely override other considerations, in particular the process of niche diversification. Community dominants, no matter how convergent their form and function, never evolve precisely coinciding niches. Along the environmental gradients which pertain as we have seen in most ecosystems, first one dominant of the community and then another is most favored by changing levels of the particular factors involved.

Alpha and Beta Diversity

Selection pressures operating along a gradient will cause both *niche diversification* and *habitat diversification*. The first occurs as a result of competition between species in more favorable environments. As a result of this competition, variation in individual species becomes more limited, producing what is sometimes referred to as 'species packing.' This is *alpha diversity* (MacArthur, 1965). It provides increases in the total number of taxa (usually taken at the species level) encountered along particular sections of a given gradient. Alpha diversity can thus be regarded as the relative *richness in species* of a given area. MacArthur concludes that in regard to bird species, alpha diversity is very similar in tropical and temperate climates, and is correlated with community structure (Fig. 7-5).

By contrast, species can also be selected for an increasingly narrow range of tolerance for environmental factors. This gives rise to what is described as *beta diversity*. Birds and many other animals show a characteristic increase in beta diversity along a latitudinal gradient from the polar to the tropical regions. Other broad ecoclines show similar variations in diversity (Fig. 7-6). The extent of beta diversity is expressed by such measures as the *coefficient of community* and the *percentage similarity* (Whittaker, 1970). These may be said to represent the amount of *habitat diversification* (Fig. 7-7).

Ecoclines

Community change along a gradient thus has several components. There is the change in population representation expressed as alpha or beta diversity. There is the gradient of the abiotic complex itself. The precise operative factors of this complex and their relative importance are usually difficult to identify without extensive experimentation. The single factor of *altitude,* for example, may be subdivided into the effects of changes in temperature, insolation and light composition, photoperiod, wind exposure, distribution, amount, and nature of precipitation, or oxygen pressures.

It is usual to combine the biotic and abiotic elements of gradient analysis into the concept of the *ecocline* as the total biological and physiochemical variation along a gradient. Many ecoclines have been described (Beard, 1955). These follow for example a temperature gradient from low to high latitudes, a height gradient from low to high altitudes, or an exposure gradient from low- to high-tide shore levels. When such ecoclines proceed from a favorable to a less favorable environment, they can usually be correlated with a parallel reduction in productivity, in

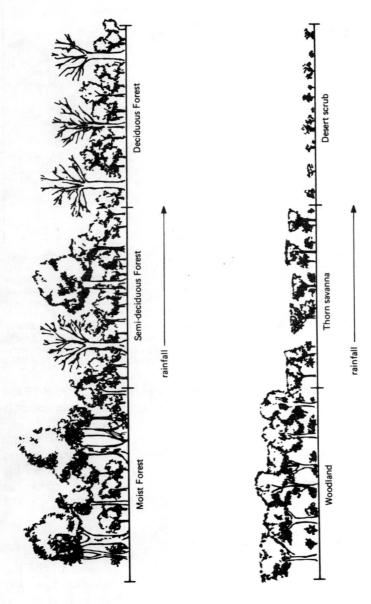

Fig. 7-6. An ecocline across a moisture gradient. From the coastal regions of West Africa with a mean annual precipitation exceeding 1500 mm., rainfall continuously declines to the desert interior, where several years may elapse without any rain at all. Along this moisture gradient there is an ecocline illustrated here by the variation in the vertical structuring of plant community types.

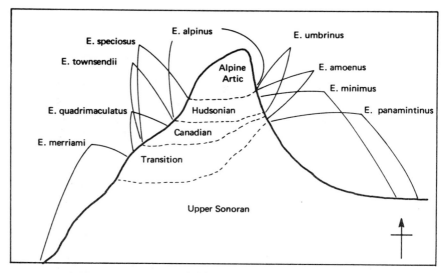

Fig. 7-7. Latitudinal gradients of diversity arise essentially from habitat diversification. Various species of chipmunk *(Eutamias)* can coexist in the same area of the central Sierra Nevada because the genus has been selected for habitat differentiation. Each chipmunk species has adapted to a somewhat different set of environmental parameters. "Species packing" in the tropics is believed to result from the progressively greater predictability of progressively warmer climates. This produces more competition and hence more niche diversification, giving a large alpha diversity. A latitudinal increase in species diversity from the poles to the equator results in some groups from both alpha diversity and from the kind of habitat diversification shown here, beta diversity. (Redrawn: based on Ingles, 1947.)

species diversity and in community structure and complexity. Thus tundra is less productive, contains fewer species and is far less structurally complex than a tropical rain forest.

A simpler kind of ecocline is illustrated in Fig. 7-8. Under the term *catena,* which emphasizes *soil* gradients, ecoclines have long been studied in the tropics (Fig. 7-8). They have been especially useful in illustrating the effects on natural communities of human occupation of particular gradients (Fig. 7-9).

Succession and Diversity

The aspects of community species composition discussed in the last few paragraphs relate especially to climax or stable communities. The situation is not precisely similar when we are dealing with seral successional stages. As already noted, pioneer communities tend to be less complex and composed of fewer populations than climax communities. The fewer plant populations which are present in the community the

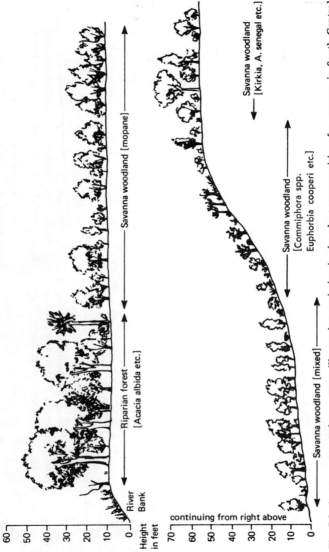

Fig. 7-8. A catena vegetation pattern illustrating variation in the plant communities of a savanna in South Central Africa. In this relatively undisturbed area plant community types conform to a topographical pattern related to the depth of soil and the availability of water there. The vegetation catena conforms closely to the soil catena pattern which determines it.

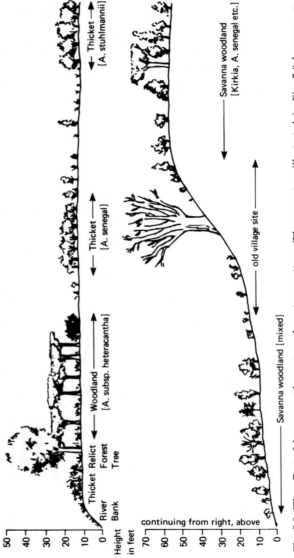

Fig. 7-9. The effect of human settlement on the catena pattern. The ecosystem illustrated in Fig. 7-8 becomes extensively degraded following prolonged human occupation. The catena pattern of plant communities which develops as a result of this degradation nevertheless still relates to the now also degraded soil catena pattern.

fewer will be animal populations that have been able to develop in the consumer trophic levels dependent on them. The climax community, with its maximized alpha diversity provides opportunities for greater niche differentiation of animals and therefore for the development of beta diversity than in a less diversified community.

Rather similar relationships develop in relation to *island communities*. Island colonization has been extensively studied by MacArthur and his co-workers (MacArthur and Wilson, 1963), who have advanced a number of theories as to the composition of the communities which occur in such situations (Fig. 7-10). Much of the attention of conservationists is focussed on the preservation of unique island ecosystems. Cities too are islands; so are centers of food production.

Stability and Diversity

The various feedback mechanisms which develop in ecosystems and which regulate population densities within communities and tend toward

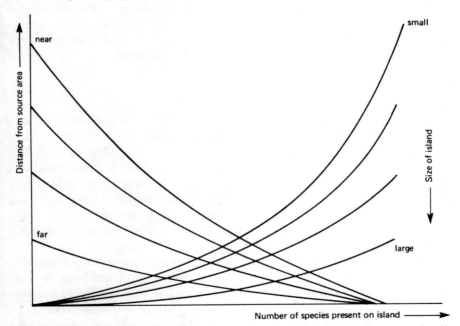

Fig. 7-10. Island colonization and species diversity is related to such factors as the distance from the region supplying the immigrant species, the habitat diversity (beta diversity) and the time which has been available for niche differentiation (alpha diversity). In this model, descending curves represent the rate of species immigration; ascending curves indicate species extinction rates. Where an ascending and descending curve intersect, immigrant species can only be established on the island if an existing species passes to extinction. (Redrawn: based on MacArthur and Wilson, 1963.)

achieving a maximum productivity in terms of the energy input into the system, are closely correlated with diversity. The greater the diversity of the community the greater its stability. The effect of human interference with ecosystems is universally to disturb communities in such a way that they lose diversity and therefore stability.

The extent to which human occupation destroys diversity depends on the extent of utilization of any particular ecosystem. The maximum destruction occurs as a result of urban development. Few if any communities can survive such drastic alteration. Such animal and plant populations as are encountered in urban ecosystems are not usually survivors from the natural communities of the area, but more commonly recent introductions of populations with characteristics which enable them to adapt quickly to the changed environment.

The *urban revolution* in which our own population created an entirely new series of ecosystems and component ecoclines, represents the most significant development in all human cultural evolution. It is barely 10,000 years old, which is only a moment of time in the three to four million years during which our genus *Homo* has been evolving. If we regard human societies as communities rather than populations, then many parallels can be drawn between society evolution and the concept of *succession,* which has already been outlined.

Succession

The main features of succession were reviewed in Chapter 2, when it was noted that it represents an ordered and progressive change in the interacting biotic and abiotic elements of an ecosystem. The first results in the replacement of one population by another, in a *time gradient* within individual communities in which species representation changes continuously and directionally. Changes in the abiotic element may be the result of these time changes in species representation, or they may result from abiotic or biotic interference originating outside particular communities.

While such theoretical analyses of the succession concept are essential, the general phenomenon is, like the existence of ecoclines, one of the most obvious ecological features apparent to any field worker. Familiar examples of succession are the gradual silting up of a shallow lake (Fig. 7-11), reinvasion of pasture by woodland (Fig. 7-12), or the colonization of the soil surface exposed in a landslip.

Plant ecologists a generation ago expended much energy on determination of the nature and direction of such *successional gradients.* They are indeed extremely important in many aspects of economic or

Fig. 7-11. Succession in a shallow lake. Stages in a succession from open water to closed forest. The newly formed lake is oligotrophic, essentially devoid of vegetation. In the foreground a floating peripheral mat of vegetation ("muskeg") is forming, sediments are settling beneath this. Behind this is a shrub zone rooted in deposited sediments. Further beyond is a forest similarly established on old sediments.

applied ecology such as range management and forest regeneration, not forgetting the aquatic successions which are involved in artificial lakes and dams. Succession may be halted short of achieving a stable or climax community by the insertion of *biotic* (e.g., fire, grazing) or *edaphic* (e.g., drainage, flooding) factors. The steady-state balance in such communities is then readily disturbed by relatively minor variations in the level at which such factors operate. This is in marked contrast with the stability of the true climatic climax, which is very stable and may even persist for a time when conditions become less favorable for the dominant species of its communities (a postclimax stage).

Considerations of postclimaxes, and longer-term successions such as those on landslips, lava flows, or similar newly exposed surfaces, have an extended time element which eventually requires a consideration of succession in terms of geological time. This aspect of succession was also discussed in Chapter 2, together with the role which our own now dominant species plays in this (Figs. 7-13 and 7-14).

This brief review of ecological fundamentals, concepts, and principles presented in these chapters concludes now with a final statement relating especially to the future of our own species in the urban ecosystems which we now find essential for our continuing support and evolution.

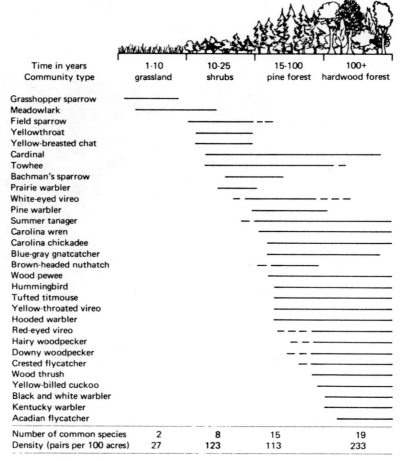

Time in years Community type	1-10 grassland	10-25 shrubs	15-100 pine forest	100+ hardwood forest
Grasshopper sparrow				
Meadowlark				
Field sparrow				
Yellowthroat				
Yellow-breasted chat				
Cardinal				
Towhee				
Bachman's sparrow				
Prairie warbler				
White-eyed vireo				
Pine warbler				
Summer tanager				
Carolina wren				
Carolina chickadee				
Blue-gray gnatcatcher				
Brown-headed nuthatch				
Wood pewee				
Hummingbird				
Tufted titmouse				
Yellow-throated vireo				
Hooded warbler				
Red-eyed vireo				
Hairy woodpecker				
Downy woodpecker				
Crested flycatcher				
Wood thrush				
Yellow-billed cuckoo				
Black and white warbler				
Kentucky warbler				
Acadian flycatcher				
Number of common species	2	8	15	19
Density (pairs per 100 acres)	27	123	113	233

Fig. 7-12. Secondary succession on abandoned farmland in Southeastern United States. The occurrence and frequency lines for each passerine bird species represent estimated breeding pairs per 100 acres. There is an increase in alpha diversity as habitat diversification proceeds through the succession from grassland to hardwood forest, over a time span of a century or more. (Modified from data of Johnston and Odum, 1956; and redrawn from Odum, 1959; reproduced with permission of Holt, Rinehart and Winston, Inc.)

Conclusion—Human Ecology

From this brief survey of ecological fundamentals supplied in this text it will be apparent that ecology, as the discipline concerned with the nature and functions of communities in relation to their environment, has indeed become an independent and unified subject. The con-

Fig. 7-13. Colonization of open water—aquatic situations provide numerous examples of pioneer communities, as shown in this photograph of Lake Kariba, on the Zambesi river between Zambia and Rhodesia. The construction of a dam in 1958 provided a man-made lake whose immediate off-shore waters were rapidly colonized by floating masses of water fern (*Salvinia*) on which dense clumps of other pioneer plant species became established.

cept which made this possible is that of the ecosystem, and the development which most insistently demands a continuity of this common theme is the introduction into ecology of the technique of systems analysis.

The systems approach, with the essential support of automated data processing provided by digital computers, ensures that the divisive tendency which previously separated autecological and synecological investigations is no longer operative. Coupled with major advances in telemetering and satellite sensing, we now have the technical power to analyze complex situations which previously defied mathematical analysis because of the massive computational involvement.

That is not to deny that autecological studies can continue. Indeed they are greatly promoted by the systems approach and the use of automated instrumentation. We may anticipate the publication of many detailed investigations on evolutionary and environmental interactions as a consequence. In the next few years the greatest emphasis will however lie in the areas of *productivity, resource exploitation,* and *recycling.* Ecosystem models providing information on nutrient and energy budgets, on primary and secondary productivity, on species diversity, and on

Fig. 7-14. Little if any tropical rain forest in the world represents a true "climax community"—more usually it constitutes a mosaic of secondary successions on abandoned cultivation clearings. Human populations have occupied most tropical rain forest areas for at least four or five thousand years, dividing it up into territories. Each group has worked over its own territory on a "swidden agriculture" or shifting cultivation pattern.

resource utilization, will be devised and improved. This will require the coordinated efforts of many workers and the kind of organizations now being assembled for the IBP biome investigations. The need for one particular type of synecological study and one kind of autecological investigation is moreover now paramount. We must have synecological information on our urban ecosystems, it is essential to instigate extensive studies of the autecology of our own species.

These two areas together comprise the specialist subject which is coming to be known as *human ecology*. Until now this subject has essentially been neglected. We liked to believe we were unique among animals, and that this conferred an immunity from the major ecological influences that affected other animals. We liked to pretend that our resources as well as our resourcefulness were unlimited. We now know both suppositions to be entirely unwarranted. Our differences from other animals are essentially *quantitative* rather than *qualitative,* a matter of degree rather than *kind*. We are turning increasingly to other animals, and especially to other social terrestrial primates, in an endeavor to discover explanations for our own behavior, which exhibits many puzzling features several generations of psychologists and sociologists have been unable fully to explain.

Failure to understand the ecological basis of our urban ecosystems has presented us suddenly and abruptly with a whole spate of environmental confrontations. Our species is now committed, by cultural adaptation, to these urban ecosystems (Table 7-1). We are no longer hunter-

TABLE 7-1

The largest cities of the world—a list of the twenty-five cities which have now exceeded a population of four million. Figures shown represent estimated thousand head of population as of 1970. The United States, with only the fourth largest population in the world, has no less than six cities in this category. This is twice as many as the country with the next largest number, mainland China, which has four times the population. The extreme concentration of the U.S. population in great cities emphasizes the highly industrialized nature of American life, and partly accounts for its present violent environmental confrontations arising from resource exploitation. The figures shown here are estimates which include not only the population within each city boundary, but also that of the commuter field in the surrounding satellite towns and suburbs.

Population in Thousands			
New York (U.S.A.)	16,077	Ruhr Cities (W. Germany)	6,789
Tokyo (Japan)	12,119	Cairo (Egypt)	5,600
London (U.K.)	11,544	Bombay (India)	5,100
Los Angeles (U.S.A.)	9,474	Seoul (Korea)	4,661
Buenos Aires (Argentina)	9,400	Tientsin (China)	4,500
Paris (France)	8,714	Djakarta (Indonesia)	4,500
Shanghai (China)	8,500	San Francisco Bay	
Sao Paulo (Brazil)	8,405	Cities (U.S.A.)	4,490
Peking (China)	8,000	Detroit (U.S.A.)	4,447
Calcutta (India)	7,350	Philadelphia (U.S.A.)	4,355
Rio de Janeiro (Brazil)	7,213	Wu-han (China)	4,250
Moscow (U.S.S.R.)	7,000	Hong Kong (Hong Kong)	4,105
Chicago (U.S.A.)	6,983	Manila (Philippines)	4,100

gatherers, evolving slowly by small genetically controlled adaptations. Our urban ecosystems provide us with the cultural information which it is essential for each one of us within our own lifetimes to exchange, learn, store, retrieve, and augment. Our unique power lies in our outstanding ability to modify the parameters of the ecological niches we occupy, instead of adapting genetically to them from one generation to the next. We have now to design our urban ecosystems so that they provide efficient, stable, and diverse systems, permitting full scope for the development of our egocentric and altruistic behavior patterns in self-perpetuating systems whose resources are internally recycled.

In order to do so we have simultaneously to solve the various environmental crises which now threaten to overwhelm not only the urban concentrations essentially responsible for them, but the whole biotic element of our ecosphere. We have perhaps no more than thirty years to do so; many ecologists indeed believe we may actually have *as little as five.* Just how long we have and why this is so is the purpose of this series to relate. This base text has surveyed the whole field of ecology. It has placed an emphasis on those ideas which it is necessary to understand in order to comprehend the fundamental basis of particular environmental problems. We now already possess much of the fundamental knowledge needed to provide a rational approach for environmental action once we have engineered the technical tools. What we conspicuously lack is an ecological understanding of our own behavior and reactions. No exposition of the nature and control of particular environmental crises is sufficient in itself; we must have the *will* and *determination* to undertake the necessary action. This base text will have served its purpose if it conveys a realization of this need for urgent, effective, and appropriate measures.

Bibliography

References

Beard, J. S., "The Classification of Tropical American Vegetation-types," *Ecology* 36:89–100, 1955.

Buell, M. F., A. N. Langford, D. W. Davidson, and L. F. Ohman, "The Upland Forest Continuum in Northern New Jersey," *Ecology* 47:416–432, 1966.

Curtis, J. T., *The Vegetation of Wisconsin: an Ordination of Plant Communities.* Madison: University of Wisconsin, 1959.

Dunbar, M. J., "The Evolution of Stability in Marine Environments: Natural Selection at the Level of the Ecosystem," *American Naturalist* 94:129–136, 1960.

Ehrlich, P. R., and R W.. Holm, "Patterns and Populations," *Science* 137:652–657, 1962.

Gleason, H. A., "The Individualistic Concept of the Plant Association," *American Midland Naturalist* 21:92–110, 1939.

MacArthur, R. H., "Patterns of Species Diversity," *Biological Reviews,* 40:510–533, 1965.

———, and E. O. Wilson, "An Equilibrium Theory of Insular Zoogeography," *Evolution* 17:373–387, 1963.

McIntosh, R. P., "Plant Communities," *Science* 128:115–120, 1958.

Paine, R. T., "Food Web Complexity and Species Diversity," *American Naturalist* 100:65–76, 1966.

Pianka, E. R., "Latitudinal Gradient in Species Diversity: a Review of Concepts," *American Naturalist* 100:33–46, 1966.

Sanders, H. L., "Marine Benthic Diversity: a Comparative Study," *American Naturalist* 102:243–282, 1968.

Slobodkin, L. B., and H. L. Sanders, "On the Contribution of Environmental Predictability to Species Diversity," in *Diversity and Stability in Ecological Systems,* Brookhaven Symposia in Biol. No. 22:82–95, 1969.

Welch, J. R., "Observations on Deciduous Woodland in the Eastern Province of Tanganyika," *J. of Ecology* 48:557–573, 1960.

Whittaker, R. H., "Dominance and Diversity in Land Plant Communities," *Science* 147:250–260, 1965.

———, "Gradient Analysis of Vegetation," *Biological Reviews* 42:429–452, 1967.

———, "Evolution of Diversity in Plant Communities," in *Diversity and Stability in Ecological Systems,* Brookhaven Symposia in Biol. No. 22:178–196, 1969.

———, *Communities and Ecosystems.* New York: Macmillan, 1970.

———, and W. A. Niering, "Vegetation of the Santa Catalina Mountains. Arizona," *Ecology* 46:429–452, 1965.

Whittaker, R. H., and Woodwell, G. M., "Structure, Production, and Diversity of the Oak-pine Forest at Brokhaven, New York," *J. of Ecology* 57:157–176, 1969.

Further Readings in Human Ecology

Brown, L. R., "The World Outlook for Conventional Agriculture," *Science* 158: 604–611, 1967.

Caldwell, L. K., "Health and Homeostasis as Social Concepts: an Exploratory Essay," in *Diversity and Stability in Ecological Systems,* Brookhaven Symposia in Biology No. 22:206–223, 1969.

Darling, F. F., "Conservation and Ecological Theory," *Journal of Ecol.* 52 (suppl.): 39–45, 1964.

Heiser, C. R., "Some Considerations of Early Plant Domestication," *Bioscience* 19:228–231, 1969.

Leopold, A., "The Conservation Ethic," *J. of Forestry,* 31:634–643, 1933.

Pimental, D., "Species Diversity and Insect Population Outbreaks," *Ann. Ent. Soc. Amer.* 54:76–86, 1961.

Spilhaus, A., "The Experimental City," *Science* 159:710–715, 1968.

Stone, E. C., "Preserving Vegetation in Parks and Wilderness," *Science* 150:1261–1267, 1965.

Review Questions

1. What is gradient analysis? Discuss the conclusions from field observations on gradient analysis so far as theoretical community ecology is concerned.

2. Explain the term *species importance.* How would you determine species importance values in a given community? Describe the attributes you would expect to characterize the populations of high species importance values.

3. Discuss the representation of niche space. What significance does this have in relation to importance values?

4. What is meant by alpha and beta species diversity? Provide named examples of each and discuss the difference between them.

5. What is the ecological significance of variations in the vertical structure of communities?

6. Discuss the relationship between succession and diversity.

7. Assume there are two opposing viewpoints, one favoring the exclusive existence of *community gradients* the other of *community mosaics*. What arguments can be produced favoring the one or the other theory?

Glossary

abiotic—nonbiological

abiotic element—a physical or chemical feature of an environment or ecosystem

acclimation—the process during which an individual organism undergoes morphological and/or physiological adaptation to one or more abiotic elements

accommodation—the location of a population within a specific area or volume of habitat

adaptive radiation—the evolution of many diverse forms within a single biotic group, each adapted to a particular ecological niche

adjustment—amendment of the accommodation of a population resulting from learning experiences

allelochemistry—the study of secondary biotic substances involved in interference occurring between two or more populations

allelopathy—interference between populations involving the release of inhibiting chemicals.

allopatric—populations with separate dispersal areas

alpha diversity—the range of populations resulting from diversification into a variety of ecological niches

altruistic—term applied to traits which have been selected because of their effects on group rather than individual survival

ammonification—the reduction of nitrates and nitrites to ammonium compounds by soil saprobes

antibiotic—a secondary biotic substance secreted by an organism which inhibits growth in other organisms

association—a conceptual grouping of populations in a community characterized by particular dominant species

autotrophic—applied to organisms which synthesize organic from inorganic substances

beta diversity—the range of populations which arises from habitat differentiation

behavior—the manner in which living systems respond to stimulation

bioenergetics—considerations of energy flow in living systems

biogeochemical cycling—the pathways of nutrients through ecosystems

biomass—the standing crop, that is the total amount of living organic material in a given ecosystem

biome—a conceptual community category comprised of vegetation characterized by a particular dominant life form, and by the associated animal dominants

biosphere—the portion of the earth in which living systems are encountered

biosystematics—the study of the biology of populations especially in regard to their breeding systems and reproductive behavior

biotic—pertaining to living systems rather than physical and chemical factors of the environment

biotic element—the organisms, populations and communities of an ecosystem

biotic potential—the maximum capacity for growth exhibited by a population

boundary effect—the results of energy exchange at an interface between a living and an abiotic system

carnivore—an animal which requires another animal as a source of food

carrying capacity—the steady state population density of a given habitat for a particular species

chlorophyll—the composite green pigment of plants

climax—the final stage of a community succession

climograph—a two-dimensional figure expressing the tolerance requirements of a particular population for two environmental parameters

co-evolution—the selection of two or more interacting populations in terms of adaptations favoring their continuing joint survival

community—a grouping of interacting populations in a particular habitat

competition—interference between two populations at the same trophic level

consumer—a heterotrophic population in an ecosystem which is utilizing dead or living organic matter as a source of food

convergent evolution—the occurrence of a number of populations of quite similar morphology and physiology as a result of adaptation in response to the same environmental selection pressures

courtship display—ritualistic behavior of an animal which preceeds mating

cytotype—a member of a population composed of individuals with essentially similar karyotypes, which differ from those of individuals in other populations

deciduous—used of structures which are shed as a result of physiological stimuli rather than obsolescence, especially leaves and teeth

decomposer—heterotrophic organisms in an ecosystem which obtain energy from the breakdown of dead organic matter to more simple substances

deme—a segment of a population in which there is no natural barrier to interbreeding

demography—the study of population dynamics, a term often but not correctly restricted to human populations

demographic transition—the voluntary reduction of the birth rate in a human population which has experienced a lowering of the mortality rate

density dependent—applied to population regulatory mechanisms which are controlled by the size of the population

density independent—applied to population regulatory mechanisms which are *not* conditioned by the size of the population

diversification—an increase in the variation exhibited by a population or community

dominance—(ecological) the phenomena in which energy flowing through an ecosystem is directed especially through a limited number of populations

dominance—(social) the psychological imposition of a hierarchial order in a population which determines the priority of access of individuals to essential requirements

dominance clique—a group of individuals in a population which has established itself in the highest position in the social hierarchy

dominant—a population which is characterized by the possession of ecological dominance in a given community

doubling time—used in a demographic sense to indicate the period which is estimated will pass before population density doubles

ecocline—directional variation in characteristics of the population or community along an ecological gradient

ecological equivalent—a population with similar adaptations but different origins which has evolved under comparable environmental selection pressures

ecological niche—a total expression of the environmental factors to which a particular population is exposed (feeding level / role + habitat)

ecological pyramid—a grouping of the successively diminishing trophic levels of an ecosystem

ecosphere—that portion of the earth which includes the biosphere and all the ecological factors which operate on the living organisms it contains

ecosystem—a conceptual unit formed from a defined series of interacting communities and all the environmental factors which operate upon them

ecotype—a segment of a population showing restricted variation in response to selection pressures arising from a particular grouping of ecological factors

edaphic—term applied to environmental factors which relate to soil characteristics

egocentric—traits which favor the survival of the individual organism

energy budget—a listing of the amounts of energy in relation to the several trophic levels of the populations in an ecosystem

endothermic—applied to homeotherms, that is animals which maintain a constant temperature more particularly by physiological processes under internal control

emigration—the movement of individuals *out* of the population

environment—the sum total of external factors to which a living system is exposed, including both biotic and abiotic ones

environmental resistance—the restriction of population growth through the interaction of one or more environmental factors

ethology—the study of animal behavior

eutherian—the term applied to true mammals, that is those which have the full development of a placenta, and bear relatively long-term embryos as a result

eutrophic—applied to stretches of water which are rich in minerals

eutrophication—the process of evolution in natural stretches of water during which the nutrient content gradually increases. Now very frequently applied to the accelerated evolution which results from the pollution of natural waters through large-scale addition of such nutrients

evapotranspiration—the total loss of water from a land surface, including both that lost from living organisms and from surface evaporation

evolution—the change in the characteristics of a population resulting from the selection of variants which differ from the mean of the original population

exothermic—applied to animals which achieve any temperature regulation they do possess through behavioral adjustment to external environmental conditions, rather than depending on internal physiological controls

extinction—the disappearance of a population as a result of the total failure of any individuals within it to reproduce the unique genotypes which it contained

family—applied in the taxonomic sense, a commonly utilized classificatory category which incorporates a range of genera possessing a number of characteristics in common

following—used in a psychological sense to indicate the association of one individual with another, usually as a result of imprinting in the earlier stages of independent existence

food chain—the succession of populations through which energy flows in an ecosystem as a result of consumer-consumed relationships

food web—a complex scheme incorporating food chain relationships between populations at various trophic levels in an ecosystem

formation—term used for a classificatory category of vegetation characterized by dominants of a specific life form

gene frequency—the ratio of the occurrence of one allele of a gene in the population in relation to other alleles of this same gene

genome—the total gene complement of an individual organism

genotypic—applied to variation which arises in an individual as the result of its possession of a specific genome

genus—a taxonomic category which represents a hypothetical assemblage of species populations having a number of characteristics in common

gradient analysis—a technique for the field analysis of ecoclines more especially in vegetation

gross productivity—the rate at which energy is procured by a particular trophic level or levels in an ecosystem

habitat—a physical portion of the environment over which a particular population is dispersed

herbivore—a heterotroph which obtains energy from the consumption of usually living plants

heterotrophic—used of an organism which obtains energy from the breakdown of complex organic substances

homeostasis—the balanced condition of a biological process in which there is no change in the final products of a particular reaction

homeothermous—the maintenance of a steady body temperature in living organisms more especially by the operation of internal physiological mechanisms

home range—the area over which an animal generally moves in obtaining its food

humification—the microbial breakdown of dead organic matter in the soil to form the largely inert product *humus*

hydrological cycle—the circulation of water through the drainage basins on which precipitation falls and eventually back to the atmosphere

hypothalamus—basal portion of the brain which controls the autonomic nervous system

immigration—the migration of individuals *into* a population

imprinting—learning to identify particular characteristics of another individual or individuals as indicating a companion relationship

interaction—the phenomena which occurs when individuals of sympatric populations encounter one another

interference—an interaction detrimental to one or more competing sympatric populations

introgression—the introduction of new alleles into the gene complex of a given population through hybridization of individuals of that population with other individuals from related populations

irritability—the response of living systems to external stimuli

karyotype—the chromosome complement of the nuclei of individual organisms within a population

lapse time—the initial portion of a population growth curve, during which increase is comparatively slow.

learning—the acquisition of behavioral patterns by an organism other than by the inheritance of direct genic controls

logistic curve—as applied to populations, an S-shaped curve of population growth which is initially slow, steepens, and then flattens out at an asymptote determined by the carrying capacity

maintenance—the continuation of a species at a particular population density in a given habitat by reproductive behavior

mineralization—the microbial breakdown of humus and other organic material in soil to inorganic substances

mortality—the rate of removal of individuals from a population by death

mutation—a transmissable change in the structure of a gene or chromosome

natality—the rate of addition of new individuals to a population by birth

natural increase—the rate of population growth as determined by subtracting mortality from natality

net productivity—the increase in energy content of an ecosystem after deducting the amount lost in respiration at all trophic levels

niche—see ecological niche

nitrification—microbial conversion of ammonium and nitrite compounds to nitrates, generally by soil nitrifying bacteria

nucleated territory—a territory which is defended with increasing vigor approaching its central point

nutrient budget—an estimate setting out for a particular living system the amounts of essential mineral nutrients which are taken up or lost

oligotrophic—applied to a body of water which contains relative low amounts of nutrients

order—a commonly used population classification category which in general usage is often employed to cluster together similar families

parasitism—a consumer-consumed relationship of two populations in which individuals of the population at the higher trophic level are usually very much smaller than those at the lower

phenotypic—pertaining to the characteristics of the mature individual organism, which is the result of interaction during development between the genotype and the environment

photoperiod—the duration of the daylight period at a particular time of year

photosynthesis—the chemical process during which green plants convert carbon dioxide to organic food substances

poikilothermous—applied to animals which achieve body temperature regulation especially through adjustment to external environmental conditions, rather than resorting to internal physiological controls

polymorphism—the existence of several states of the same character in one population

population density—the number of individuals of a population in a given area or space

population—an aggregation of similar individuals in a continuous area which contains no potential breeding barriers

population dynamics—the study of variations in population density

predation—a consumer-consumed relationship in which the prey at a lower trophic level is usually smaller than the predator at a higher level

predictability—the extent to which variation in the environment conforms to a predictable pattern

primary productivity—the rate at which energy is taken into an ecosystem through the activity of producers

producer—autotrophic populations, usually of green plants, which procure energy from outside an ecosystem and direct it into the system

productivity—the rate of procurement of energy

random drift—a directional change in gene frequency in a population which is not a response to selection pressure

recombination—the reassortment of characters within linkage groups as a result of the crossing over which occurs during meiosis in reproductive tissue

reducer—heterotrophic individual which utilizes the chemical energy of organic matter while breaking it down to more simple substances

regulation—as applied to population dynamics, the control of population density

secondary productivity—the procurement of energy by heterotrophs

sedimentary cycle—the circulation of nutrients in an ecosystem which involves geological weathering and erosion with the eventual recovery of the elements by the uplift of marine sediments to form land masses

selection—differential reproduction in individuals of a population arising from a variation in the individual optima and tolerance limits

shifting cultivation—applied to the cultivation of areas of tropical woody vegetation by clearing and the establishment of temporary farmlands, soon abandoned

social rank—the order of individuals within a population in a dominance hierarchy

speciation—process of formation of a new population sufficiently distinct from the parent to be recognized as different species

species—a group of similar individuals having a common origin and a continuous breeding system

species packing—extensive diversification into species populations within a comparatively narrow range of variation

sub-species—a geographical segment of a species population possessing a range of variation representing only a portion of the total variation of the species

succession—the replacement of one community by another as a result of changes in the environment

survivorship—the number of individuals in a population persisting to a particular age

symbiosis—an interaction between two or more populations which is favorably selected

sympatric—populations dispersed over a common area

systematics—a study of the diversity of living organisms

taxon—a classificatory unit in taxonomy which is not identified with any particular category

taxonomy—the study of the methods of classifying organisms

teleoclimate—the micro-climate at the boundary between living organisms and the environment

terpenes—a group of volatile aromatic organic substances commonly released from the shoots of flowering plants in particular families

territoriality—the identification of an individual organism, population or community with a particular spatial area or volume

trophic level—a particular step occupied by a population in the process of energy transfer within an ecosystem

xeromorphic—plant characters which appear to restrict water loss during adverse conditions

xerophyte—a plant possessing xeromorphic characters

Index